Fish Business Management

Strategy – Marketing – Development

Andrew Palfreman
Institute of Food Health Quality
University of Hull

Fishing News Books
An imprint of Blackwell Science

b

Blackwell
Science

Copyright © 1999 by
Fishing News Books
A division of Blackwell Science Ltd
Editorial Offices:
Osney Mead, Oxford OX2 0EL
25 John Street, London WC1N 2BL
23 Ainslie Place, Edinburgh EH3 6AJ
350 Main Street, Malden
  MA 02148 5018, USA
54 University Street, Carlton
  Victoria 3053, Australia
10, rue Casimir Delavigne
  75006 Paris, France

Other Editorial Offices:

Blackwell Wissenschafts-Verlag GmbH
Kurfürstendamm 57
10707 Berlin, Germany

Blackwell Science KK
MG Kodenmacho Building
7–10 Kodenmacho Nihombashi
Chuo-ku, Tokyo 104, Japan

First published 1999

Set in 10/13 pt Palatino
by DP Photosetting, Aylesbury, Bucks
Printed and bound in Great Britain by
MPG Books Ltd, Bodmin, Cornwall

DISTRIBUTORS

Marston Book Services Ltd
PO Box 269
Abingdon
Oxon OX14 4YN
(*Orders:* Tel: 01865 206206
          Fax: 01865 721205
          Telex: 83355 MEDBOK G)

USA
Blackwell Science, Inc.
Commerce Place
350 Main Street
Malden, MA 02148 5018
(*Orders:* Tel: 800 759 6102
                781 388 8250
          Fax: 781 388 8255)

Canada
Login Brothers Book Company
324 Saulteaux Crescent
Winnipeg, Manitoba R3J 3T2
(*Orders:* Tel: 204 837 2987
          Fax: 204 837 3116)

Australia
Blackwell Science Pty Ltd
54 University Street
Carlton, Victoria 3053
(*Orders:* Tel: 03 9347 0300
          Fax: 03 9347 5001)

A catalogue record for this title is available from
the British Library

ISBN 0-85238-255-3

Library of Congress
Cataloging-in-Publication Data
Palfreman, Andrew.
  Fish business management:strategy,
  marketing, development/Andrew Palfreman.
    p. cm.
  Includes bibliographical references and
  index.
  ISBN 0-85238-255-3
  1. Fish trade.   2. Fish trade–
Management.   3. Fishery law and
legislation.   4. Fishery management.   I. Title.
HD9450.5.P27   1999
639.3'068–dc21   98-33448
                                         CIP

For further information on
Fishing News Books, visit our website:
http://www.blacksci.co.uk/fnb/

Contents

Preface

This book is based very largely on the personal experience of the author. It amounts to a review of issues which he has had to deal with at various points in a long career in the fish industry, much of it in the private sector or in close association with people from the private sector. The book is not a technical manual for fishing, fish processing or fish farming. Rather, it is written to draw attention to the commercial aspects of the fish industry in the widest sense. Throughout the book 'he' is used to stand for both genders.

Acknowledgements

I am grateful to the Department for International Development for permission to use a study completed for them of the fisheries sector on Hatiya Island, Bangladesh. The original study is cut and amended to bring out some key commercial features of the fisheries sector. Thanks are due to the Natural Resources Institute, as well as to the two Romanian companies, for permission to use abridged versions of business plans as examples in Chapter 7. I also wish to thank the government of Vanuatu for permission to make use of an evaluation study, adapted for Chapter 16. The text is strongly influenced by an analytical framework in John Kay's book *Foundations of Corporate Success*. This approach is used here by permission of Oxford University Press.

Andrew Palfreman

Chapter 1
Setting the Scene for Fish Business Management

Thinking about strategy

Strategy is often presented as the bundle of policies adopted by a business or a country to achieve well-defined goals (for example, Kotler, 1991, p.35). Strategy is formulated as a result of an assessment of the strengths and weaknesses of the business from the inside, and the opportunities and threats in the market place surrounding it. Undoubtedly many people have found this basic approach to be helpful and for this reason Chapters 2 and 3 discuss what strategy is and demonstrate how the approach works. This book also presents an alternative to the traditional methodology, one that has proved to be successful in practice and avoids some of the potential pitfalls of the earlier approach. John Kay, in his celebrated book *The Foundations of Corporate Success* (1993), describes the method in detail and it is adapted here for the fish industry. In this book we regard the two approaches as being complementary, but give more weight to the second because it seems to be a closer match to reality and it also makes more use of 'economic' concepts. The second approach permeates the whole of this book and this chapter supplies an introduction to it.

Distinctive capabilities

Whether you are setting up a business for the first time, trying to expand an existing business or thinking of buying one, the first question you have to ask is:

'What can I offer to the marketplace which is at least as good as, and preferably better than, the competition?'

There are four basic ways in which this can be done. The categories identified are borrowed from *Foundations of Corporate Success* (Kay, 1993):

- Architecture
  This category splits up into four:

  (1) relationships with buyers

(2)   relationships with suppliers
(3)   relationships between people who work in the business
(4)   relationships with other businesses

- Reputation
- Innovation
- Strategic assets

Let us suppose that the profit and loss account in Table 1.1 is for a skipper-owned fishing vessel which has moved from a modest profit to a loss between 1996 and 1997. Let us examine some of the things that its owner might do, under each of the above headings, to arrest the apparent decline.

Table 1.1 An accounting framework.

Profit and loss account	1996	1997
Turnover	300 000	250 000
Cost of sales	200 000	200 000
Gross profit	100 000	50 000
Operating expenses	75 000	75 000
Operating profit	25 000	−25 000
Net interest	1 000	1 000
Net profit before taxation	24 000	−26 000
Taxation	2 000	0
Net profit after taxation	22 000	−26 000

Architecture

Relationships with buyers

Sometimes businesses are able to establish good relationships with buyers, which gives them an advantage over the competition. Fishing skippers have been able to improve the prices they receive, compared to others, by paying greater attention to fish quality. The fish buyers get to know a particular vessel and recognise that the fish is handled with more attention; hence they bid with more enthusiasm. Similarly, if the skipper of a freezer trawler can begin to build a reputation for supplying fish closely matched to a buyer's specifications, he might then negotiate a better price. So it may be possible to strengthen the profit and loss account through focusing on fish quality. In the example the skipper can think about how he might begin to improve his turnover through better prices. Pricing strategy is one dimension of the sales process, but there are other aspects too, discussed in later chapters.

Relationships with suppliers

Some businesses establish special relationships with suppliers. This can lead to commercial advantages, such as the manager of a fish restaurant who always buys from one fish farm because he is sure of the quality of the product and the regularity of supply. Repeated transactions, and therefore a developing mutually advantageous relationship, can contribute to fishing profitability. So the fishing skipper needs to examine the structure of his inputs to see what can be done. A typical example might be the supply of technical expertise for undertaking engineering tasks. A skipper might establish an excellent personal relationship with a particular individual or company, and so get an improved service, and perhaps in the short-term some supplier credit. Thus he might be able to get his operating expenses down through better attention to the maintenance of his vessel.

Relationships between people who work in the business

Sometimes businessmen can achieve a higher degree of commitment and team spirit from the people they work with and sometimes people work exceptionally well together. This can be a source of competitive advantage. The successful fishing skipper is invariably someone who, one way or another, inspires his crew to greater degrees of effort. He might be able to increase the catch, and thus the turnover, through greater effort and thus drag the vessel back into profit.

Relationships with other businesses

Sometimes businesses can achieve a great deal by working with other businesses in the same industry. Finding ways of cooperating with each other helps to enhance the competitive advantages of both, and game theory provides a framework for analysing cooperation and competition (Miller, 1992). It is a very important theme of this book that one of the most important ways in which European Union fish producers can enhance their sales revenue (or turnover) is through joint action under the legal umbrella of the officially recognised Fish Producers' Organisation. Organising collective action without legal support is much harder but may be worth trying in many circumstances. There may also be more informal cooperative arrangements, such as sharing information, which may help to increase profitability. For example, if skippers can get to lucrative fishing grounds at lower cost through sharing information they can reduce the 'cost of sales' in the example.

Reputation

For certain types of products and services it is essential to work on building a good reputation, which can make enormous differences to chargeable fees. It is a question of improving the image of the company in the eyes of buyers. The idea is best applied to providers of services to the fishing industry, such as legal firms. They build up a reputation in the industry for supplying good quality legal advice and thus they attract more and more business.

Innovation

Businesses must innovate because innovation results in higher productivity and increased profits. The problem with innovation is that it can be copied, and will be if it results in disadvantages for competitors. Innovation cannot be avoided if a fish business is to remain level with the competition. New commercial ideas are a source of extra profit for the business, at least until competitors copy the idea.

Strategic assets

Strategic assets are those assets that do not depend on any special qualities of the people who work within the business. Examples include monopoly ownership of the lines for delivering electricity, of reservoirs and water pipes, and the right of universities to award degrees, but many businesses have them to a greater or lesser degree. For people who intend to take up a long-term career in fishing the most important strategic asset is a fishing licence; it is a source of economic added value which does not depend on special capabilities of the skipper. For a fish farmer, a water abstraction licence might be a similar strategic asset.

From distinctive capabilities to competitive advantage

Opportunity costs

An entrepreneur needs to look deep within his own business, or the business he is thinking of setting up, and to ask himself what his special abilities are. The next question is whether these distinctive capabilities can provide him with a competitive advantage in the market place. This is where economics can provide an insight because it tells businessmen that they need to look at more than accounting statements when they are analysing their business performance. They have to look at *opportunity costs*. Opportunity costs are the cost of an input in its *next best alternative use*. It is particularly important for businessmen to think about their own opportunity costs when they consider setting up a small business. A business might look profitable if expected revenues are likely to cover costs. But the owners of small businesses sometimes forget to cost themselves at their true opportunity cost. They anticipate drawing a very small amount from the business, which is substantially less than they could earn elsewhere, working for someone else.

Businessmen also underestimate other costs when they are thinking of setting up a business. A common mistake is to fail to take into account the cost of capital, especially if it is the entrepreneur's own money that is going into the business. He should really calculate this by looking at how much the capital would earn in its next best alternative use. That is the true cost of capital he puts into the business. The fish farmer needs to consider how much the land he puts into fish farming would earn in its next best alternative. The enthusiast is sometimes tempted to underestimate the true opportunity costs of the land he is using. As far as other costs are concerned the rule can often be followed that market prices, or prices paid to wholesalers and retailers, are a good enough measure of opportunity costs.

Accounting for opportunity costs

When a businessman assesses a new business he should ask himself whether it covers his opportunity costs as well as his accounting costs. In fisheries there are three fundamental questions:

(1) Is the business going to generate an income for the entrepreneur which at least matches what he would earn doing something else? If not then he needs a good reason to be going into the business, such as interest or independence.
(2) Is the capital going into the business likely to earn a return at least as good as it would earn in its next best alternative? A careful investment appraisal of the business is required to establish this.
(3) Is the land which the business is planning to use going to earn a return at least as great as its next best alternative use? If the land belongs to the entrepreneur he should *impute* a rent to it. This means that even if he does not have to pay any rent to himself he should consider how much he would get if he rented out the land to someone else and should include that cost in his calculations.

When the businessman has done this, he can calculate his *unit costs*. He does this by dividing his total costs per year by the expected output. So in the example the fishing vessel's accounting costs are £276 000. Taking opportunity costs into account results in the entrepreneur putting this figure up to £280 000. Then if annual output is 500 tonnes, the unit cost of production is £280 000 divided by 500, or £560. If this is less than the unit costs of production of others then he can be said to enjoy a *competitive advantage*.

Converting competitive advantage into commercial gain

Competitive advantage is turned into *commercial gain* in the market place. Sometimes this commercial gain is termed 'economic value added' or EVA, or in the older 'economic' terminology, 'rent'. When fish businesses are able to sell their output for more than the unit opportunity costs of production we prefer to use the term economic added value, EAV, to denote the resulting surplus. This book is about the various ways in which businesses in the fish industry can identify their distinctive capabilities, secure competitive advantages and convert them into EAV.

Game theory

Game theory is a branch of applied mathematics which studies how people and businesses make decisions when they are mutually interdependent. This means that the player does not know for definite what the outcome of his strategy in the game is going to be, because that depends on someone else's strategic choice, as well as his own. Indeed, a better name for *game theory* would be the *theory of interdependent decision-making* (Colman, 1995). The language of game theory is now used very widely. Because the fishing industry has many examples of interdependent choices, especially in the catching sector, and it also provides a very useful framework for analysing many fisheries situations, this book introduces readers to it.

Summary

The theme motivating this book is that fish businessmen can become more successful if they ask themselves three fundamental questions:

(1) What are our distinctive capabilities?
(2) How can these unique capabilities be turned into competitive advantages?
(3) How can competitive advantage be translated into added value in the market place?

   The book is addressed to those charged with the development of a thriving private fisheries sector. It suggests that sector analysis is best completed with the same framework in mind. The subsequent chapters expand upon these basic themes.

Chapter 2
Strategy and the Business Mission

What is a strategy?

Strategy can be thought of as the game plan that the senior management of a company has chosen to play. It is the particular path which the company follows as it attempts to achieve its objectives and goals.

There are many different options for a fish business. For example, a catching business may adopt a low cost, low investment strategy. It may foresee that there are limited opportunities in the future so it opts to run its capital down and diversify. It may, therefore, adopt a strategy of capital disposals in order to obtain cash which it can then use for new investments outside the fish industry. Alternatively it may decide that there are likely to be opportunities for efficient companies, so it may choose to upgrade its catching capacity, investing in modern vessels and fishing techniques.

A fish processing company may adopt the strategy of becoming the lowest cost producer of a product which is, to a greater or lesser degree, a commodity (such as fish fingers). Alternatively it may aim to be highly flexible, with short production runs, producing more differentiated products which can attract the higher prices needed to compensate for the implied higher production costs. A company may have a strategy pointing towards diversification away from fish products towards related products. A typical strategic issue for an eastern European fish farm business might be whether it should continue to use land for extensive fish farming or try some other commodity. It might also try to vary the species-mix or the degree of intensity of fish farming.

Strategy also concerns the financial side of business. For example, the financial strategy may be to move from being a private limited company, in which share-holding is concentrated in the hands of a few members of a family, to being a public company with a wider ownership of publicly traded shares. An alternative strategy for a public company in decline may be to revert to private status under which the original owner buys back the shares from the wider share ownership.

Three examples of strategic planning and management

Three different examples of strategic planning and management are given to illustrate its fundamental role in the functioning of any organisation. For Dairy

Crest the issue is from whom should the company be buying milk. Albert Fisher wants to divest itself of its least profitable division to concentrate on more successful lines. The State of Queensland wants to nurture a small but prosperous and environmentally friendly aquaculture industry.

Dairy Crest

Financial Times (5 November 1997) reported that Dairy Crest, one of the three UK's biggest dairy companies, had announced that it would be forced to reconsider buying most of its milk from Milk Marque, the dominant supplier of milk in the UK. Dairy Crest said that it was awaiting action by the Office of Fair Trading against Milk Marque's selling system and move into cheese processing. Mike Dowdall, chairman, said Milk Marque's acquisition of Aeron Valley Cheese was an attempt to manipulate the milk price and 'does long-term disservice to producers (farmers) and manufacturers'. Dairy Crest buys 75% of its milk from the farmers' cooperative (the successor of the Milk Marketing Board) and is its biggest single customer. John Houliston, chief executive of Dairy Crest, wanted a 'true partnership' with Milk Marque, maximising both parties' income and eliminating waste and said that unless Milk Marque was willing to reconsider its approach, Dairy Crest's strategy would change. In response, David Yeomans, Milk Marque's chief executive, defended Milk Marque's pricing policy. Dairy Crest's strategy has been to meet its milk requirements from Milk Marque. However circumstances have changed – a higher price for milk – and this is giving the company cause for thought.

Albert Fisher

The Financial Times (*FT*) (2 November 1997) reported that the well-known food manufacturer Albert Fisher had decided to sell its seafood business, which had previously been part of its long-term plan, use some of the proceeds to buy back a chunk of its shares, and concentrate on its remaining activities, which include fresh produce and food processing. According to the *FT* the company had some difficulty in finding a buyer.

Albert Fisher's 1996 turnover by division (£m)

Seafood	396.3
North America	181.5
Europe, fresh product	390.5
Europe, processed foods	271.3

Queensland Aquaculture

This summary of the Queensland Aquaculture Industry Vision (part of the Queensland Aquaculture Development Strategy) is an example of an Australian state institution adopting a strategy for the development of the aquaculture business in its region. According to the document, Queensland seeks:

'A profitable, sustainable and diverse multisectoral aquaculture industry servicing the domestic and worldwide demand for targeted aquaculture products' (Queensland Department of Primary Industry, 1997).

The aim is that by the beginning of the next millennium Queensland will have an aquaculture industry with a value in the order of A$ 125 million, directly employing some 2000 people throughout the state in secure jobs. The international competitiveness of its products will be guaranteed by their premium quality and market demand.

The intervention of the state is envisaged through focused research and extension, appropriate management and regulation and support for effective industry associations. The specific policies are:

'the maintenance of a clean aquatic environment with low levels of contaminants and disease through sound environmental planning, design and operations;
highly competitive operations through the use of appropriate technology;
production tailored to market niches and requirements;
sector industries of an economically viable size, with sound financial planning and backing;
equitable and effective regulatory processes.' (Queensland Department of Primary Industry, 1997, p. 2–3.)

The role of senior management

A major responsibility of senior management (at director level) is setting strategy. This is much more difficult than may appear to be the case at first sight. This is because (1) the external environment in which the company is operating must be read correctly and (2) the internal strengths and weaknesses of the business must be fully understood. Consider the example of a fish processing company which has made a strategic decision towards the production of value-added fish products. Senior management must ensure that the financial resources are available to pay for the detailed and careful market research that is necessary. It must also ensure that the company has an internal project team (see Chapters 14 and 15) with the capability and time to liaise with whoever has done the research. The strategic choice also has implications for the financial resources of the company over a longer time period. It will have implications for the allocation of resources inside the business if the results of the market research are to be turned into action. To do all this thoroughly will inevitably absorb much of senior management's time and attention.

Senior management has other, related, responsibilities. It must ensure that the people who work for the company understand the implications of the chosen strategy. It must set performance indicators to check that the chosen strategy is being carried out. It must always be ready to review the way the company is going about things if events turn out unexpectedly.

The business mission

We open the discussion of strategy formulation and implementation with a description of the conventional approach to it, summarised in Table 2.1. For reasons which will become clear, whilst the traditional approach has many strengths, there are modern alternatives to it. One such approach, outlined in Chapter 1, is described in detail in Chapters 4 and 5. It is intended to complement, rather than compete with, the more traditional approach, described in this chapter and Chapter 3.

Table 2.1   The strategic planning process.

Stages	Actions	The influence of monitoring and evaluation
1	Defining the business and establishing a business mission	Based on all relevant information available to the company
2	Analysing strengths and weaknesses (internal) and opportunities and threats (external) (SWOT)	Leads to a review of the business mission (Stage 1)
3	Setting strategic objectives and targets	Review business mission, and SWOT analysis
4	Formulating a strategy to achieve objectives and targets	Review business mission, SWOT and objectives and targets
5	Preparing a business plan	Review business mission, SWOT, objectives and targets and strategy
6	Implementing a business plan	Check all previous assumptions and information
7	Monitoring and evaluation of the strategic planning process	Conscious feedback and review of entire process

A good place to start the process of defining a strategy is with a business mission. The aim of the mission statement is to give the people working in the company a sense of purpose and to project to the rest of the world how the business sees itself. The statement of the corporate mission answers questions such as:

- What business are we in?
- Who are our customers?
- What should we be doing to improve our business?

A great deal of management writing (such as reported in Ohmae, 1983; Peters and Waterman, 1982; Clifford and Cavanagh, 1985; Barrow *et al.*, 1992) places emphasis on senior management motivating the staff of the company through a coherent, thoughtfully prepared and frequently repeated corporate mission statement or statement of business purpose. The phrase 'mission statement' has recently been dropped from some corporate public material (such as the Northern Foods annual accounts), but the principle of setting down in a statement of up to about 50 words what the business is all about remains. Some institutions have gone further and used the mission statement to write a short, punchy phrase to encapsulate the values of the business. What about the Western Australian Fishing Industry Council's 'Seafood Forever!' as an example of this?

The starting point for the strategic plan is the preparation of the mission statement, or its equivalent if the company wishes to sound less evangelical. It is shaped by five elements: the history of the company, the current preferences of owners and managers, the market environment, resources available to the company and its distinctive competencies. It has to be subject to review and change as the operating environment of the business changes.

The mission statement may be understood as a general statement of company policy which is intended to make a statement to the world outside about what the business is all about and to influence the working practices of all the people within it. It is also generally recommended as the first item in a typical business plan (see Chapter 7) that might be presented in an application for a loan from a commercial bank. It gives the potential financier a first impression of the company and a summary of what the company thinks it is all about.

Some examples

A North Atlantic fishing company

On both sides of the Atlantic the fishing industry faces a number of difficulties. Despite fisheries management measures the fisheries resources of cod, haddock and other bulk demersal species have continued to diminish. At the same time the fish produced in some regions has never achieved an international reputation for high quality. The following is an imaginary mission statement for a company operating a small fleet of offshore wet fish trawlers.

'The company aims to secure a sustainable return for its shareholders. To this end it will work within the fishing vessel owners' association to persuade the government to introduce effective measures to manage the principal fisheries. These measures should be designed to secure the long-term future of the catching sector of the industry. The company also intends to work to improve the first-hand price of fish received by its vessels by improving the quality of landed fish.

The company will also work through the fishing vessel owners' association for better quality control in the fish packing plants and to secure the international recognition of fish from this region.

The company aims to increase the ownership interest of loyal and long-standing fishing skippers.'

Given the same background as for the fishing company the following is an imaginary mission statement for a fish processing company in the same region. Although this company is interested in fisheries management its emphasis is quite different. It is fighting for low-cost production so that it can improve its profitability.

'The mission of the company is the creation of sustainable profits for the shareholders.

It will work within the fish packers' association to urge the government to introduce fisheries management measures aimed at securing a sustainable supply of good quality, lower cost fish.

It will initiate a programme of product and market development. It will aim for international recognition of its brand name. It will strengthen its system of quality control and introduce HACCP throughout the production system.' [HACCP stands for Hazard Analysis Critical Control Point.]

A new shrimp culture business

The following example has an Indian flavour, and might be a typical mission statement for a new shrimp culture business on the Indian subcontinent.

'Quality Shrimps Ltd aims to achieve sustainable profits within 3 years of registration. To this end it will grow and supply high quality fresh shrimps to shrimp exporting companies.

The company will strive to maintain its fish ponds and equipment in excellent working order. It will do this by allocating adequate resources to repairs and maintenance. The company has a policy of technological innovation. It will allocate resources to research and development and will introduce such new technological developments as are profitable for its business operations.

The company will pay a fair wage to its workforce, but it will not permit a national trades union to operate within its area of competence.

The company will work within the National Shrimp Growers' Association to ensure that waste disposal standards of shrimp culture companies are improved.'

The Western Australian Fishing Industry Council (WAFIC) has a mission:

'The WAFIC Mission is to increase the economic benefit from the fishing industry to the people of Western Australia through planned programmes that will enhance the value of sustainable aquatic resources.' (Western Australian Fishing Industry Council Research, undated, p.iii.)

Northern Foods does not have a statement actually described as a mission, but its introductory remarks in its Annual Report for 1997 amount to very nearly the same thing.

'Northern Foods is a leading UK food producer, focused on two distinct operating areas: Prepared Foods and Dairy. Our Prepared Foods business is Britain's foremost supplier of high quality chilled foods under the own labels of major retailers; it also has a strong branded presence in biscuits, fresh chilled dairy products, frozen food and savoury pastry products. Our Dairy business is the country's largest supplier of liquid milk to retailers and doorstep customers. We have attained our strong market positions by combining an emphasis on quality, innovation and service with consistent investment in improved facilities and technology.' (Northern Foods plc. 1997, p.1.)

Yorkshire Electricity has a corporate mission:

'To be the leader in our chosen markets within electricity distribution, electricity supply, gas and generation, providing the best levels of customer service.' (Yorkshire Electricity Group plc., 1996, p.7.)

Main elements of a mission statement

The success of a business depends very much on how it defines itself. The story of many companies changed when the directors thought through again what business they were really in. For example, the long-established shipping business P&O defined itself as in the transportation business because that is what the company was good at doing. It then extended itself into other areas of transport. In Britain, those deep sea companies which were able to redefine their role were the ones who survived and prospered. One company built on its competence in cold storage. Others extended themselves into offshore services for oil exploration. So what should a mission statement contain?

- Customers: who are they?
- Products and services: what are they?
- Markets: where does the business compete?
- Technology: what is it?
- The future: where is the business going in terms of profitability and growth?
- Values: what are the business's basic beliefs, attitudes and priorities?
- Self-understanding: what are the business's strengths and weakness?
- Public image: what sort of impression does the business want to give of itself to the general public?
- Employees: what is the attitude of the business to its staff?

Summary

Strategy

Businesses take strategic decisions. There are invariably a variety of ways of approaching a business problem and strategy is the means by which companies choose the path to follow. They choose a strategy by examining the marketplace and making a judgement about what they should do, bearing in mind the reaction of competitors and collaborators and the capabilities of their own personnel. Strategic choice is the responsibility of senior management.

The business mission

The formulation of the business mission is an appropriate starting point in the process of developing a corporate strategy. The mission statement is a succinct description of what the business thinks it is all about. It answers questions such as:

- What business are we in?
- Who are our customers?
- What should we be doing to improve our business?

Chapter 3

The Business Environment, Strategy and the Business Plan

Analysis of the business environment

Many textbooks (for example, Curtis, 1994, p.327) approach the problem of the examination of the business environment through so-called SWOT analysis. This chapter supplies an exposition of this for completeness, and then summarises how strategies are turned into plans and projects. SWOT stands for Strengths, Weakness, Opportunities, and Threats. Strengths and weaknesses concern the interior of the business and the opportunities and threats are in the world outside the business – the external environment. Bowman (1990) acknowledged that the SWOT approach has now been superseded by new methods. Chapter 4 describes one of these newer methodologies for strategy formulation and applies it to the fish sector.

The external environment

Economic changes

It is a truism to say that the world is always in a state of flux. However, for a fish business it is important to be aware of these changes to the extent to which they impinge upon the outlook for the company. For example, it is quite surprising how little attention businesses sometimes pay to macro-economic changes. It is well known that the demand for demersal fish is often strongly related to income. The better-off people are, the more they are likely to want to buy fish and fish products. It is also true that the demand for fish is influenced by the price of substitute products, such as poultry (especially chicken). So if incomes in a market are rising (macro-economic changes), this may suggest a possible area for market development. Exchange rate changes are a crucial factor behind the profitability of different markets and expectations of changes will influence where fish businesses make their marketing efforts. Interest rate changes may influence investment policies. Expectations of higher rates of inflation may suggest that a company should be investing in physical assets, particularly land, rather than cash-related assets. Many fish businesses in the Far East should be alert to the probable difficulties, arising from economic mismanagement, that some economies are now in (1999).

Technological changes

Technological change implies that people are always finding new, and usually lower cost, ways of producing things. This takes place in the fish industry and fish business managers must always be aware of new developments. The advance of modern navigation equipment is a very clear example of a technological change with a strong influence on the catching sub-sector. Other influential changes include the increasing sophistication of vessel monitoring and surveillance systems and the use of specially designed computers as fishing logs. Aquaculture businesses need to be aware of the scope for computer-controlled production systems. An example from the fish processing side is the need for companies to maintain awareness of developments in production techniques and packaging.

Political and legal changes

Political and legal changes have an influence on business in a wide variety of respects. For example, the catching sub-sector of the fishing industry is obviously affected by legal changes relating to fishing limits, conservation of fish, factory regulations, tariffs, employment law, regulations on quality control and so on. A similar list could be drawn up for fish processors and fish farmers.

Social and cultural changes

Examples of social and cultural changes relevant to the fish industry in Western Europe and the United States include the following: changing attitudes to red meat and awareness of the low-fat characteristics of many fish species (a cultural change); the change in food consumption habits towards fewer set family meals and more 'grazing' (an important social change).

Changes in the natural environment

A fishing business has to be aware of natural changes in the environment, due to climate change or various other factors, because these will influence its strategic choices. It must also be aware of changes due to the influence of man, such as declining fish resources where there is open access, or improving fish resources, perhaps due to better fisheries management.

Competitors and customers

The second strand of external environmental analysis is to take a more narrow focus, to look in more depth at those areas more closely related to the company's markets (see Chapter 8). A business must always be aware of how its customers are changing, what its competitors are doing, changes in distribution channels and developments on the supply side.

Competitors

The fish business is global and competition is usually severe. It is also often distorted by systems of subsidies and tariffs. A fish business has to keep its eyes on the competition. For example, a fish catching company will want to know about advantageous arrangements that others may have secured on productive fishing grounds. It will want the same or better terms itself. A fish processing company will want to know if other businesses have secured access to lower cost supplies of fish than itself. A fish farm business must obtain information on the technologies that others have invested in.

Customers

Customer developments are also very important. Fish buyers have certain requirements and these change. There may be a requirement for fillets of a certain size, or fish boxed in a particular way, or handled in a special manner. Finding out precisely the requirements of markets in terms of fish presentation, quality and packaging is in itself a major exercise.

Distribution channels

Fish businesses need information about distribution channels. They need to know what are the channels for selling fish into a new market and, of these, they need to know which will be the most financially rewarding for themselves.

Suppliers

The other major aspect of the immediate business environment is suppliers. For example, fish processors and distributors need to know where their supplies are likely to come from and what steps they need to take to secure supplies of the appropriate specification and quality. The supply of inputs can also be a problem for the aquaculture sector, especially if economic changes make obtaining new equipment, feeds or other inputs difficult.

The internal environment

The strengths and weaknesses of SWOT are features of the business's internal environment. A business has various strengths. It may have financial strengths in the shape of exceptional expertise and/or resources. It may be particularly skilled at fish marketing, with detailed knowledge of export markets. It may also have exceptional engineering skills, with expertise in ship repair work. One significant competence may be its organisation. In today's highly competitive environment, having an organisational structure which responds intelligently, flexibly and fast to changes in the marketplace is a great advantage.

For example, if a fish catching business can prepare one of its vessel for some other task, such as oil rig support or hydrographic surveys, more quickly and effectively than the competition, this could be an advantage. Similarly, the ability of food processing industries to change production lines quickly in response to a very dynamic marketplace can give them a significant commercial advantage.

A business may also have a whole variety of internal weaknesses, such as too many people to operate profitably, being technologically behind the times, aging staff, assets which are out of date, and many others.

Objectives and targets

From the in-depth analysis of the business, management can begin to develop some strategic objectives. These should be realistic projections from the analysis described above. Books and articles about business planning sometimes make a distinction between 'objectives' and 'targets' and also sometimes use the word 'goal' in the place of 'target'. The convention adopted here follows the usual practice and treats 'objectives' as non-quantitative aims and 'targets' as quantitative aims. This is not only a point about definitions; it is also about the process of strategic planning. Realistic objectives for the business are developed and then these are refined and narrowed down to something much more specific and quantified which we call targets. Some typical business objectives, developed out of the strategic analysis described above, perhaps SWOT, but perhaps also influenced by an analysis of distinctive capabilities and competitive advantages, are described below.

Financial

Some characteristic financial objectives might be as follows:

- Survive and avoid bankruptcy
- Become and remain profitable
- Maximise return on investment
- Increase cash flow
- Reduce the burden of debt
- Increase dividend payout
- Avoid financial takeover.

These might be refined into the following corresponding set of targets:

- Within 12 months reduce overhead expenditure by 20%.
- In 2 years eliminate company losses and in 4 years achieve net profits of 20% of capital employed.
- Undertake capital investment only if it is capable of achieving a rate of return of 15% or above.
- Increase net cash flow from $100 000 per year to $120 000 per year by the end of 3 years.
- Reduce the gearing ratio (the ratio of debt to equity) from 50 to 30% over the next 5 years.
- Over the next 5 years increase the dividend on $1 shares by increments of $0.01 per year.
- Boost the share price by 20% in the next 3 months.

Marketing

A similar exercise might be undertaken for markets. Some examples of marketing objectives for a typical fish catching business might be as follows:

- Increase the landed value of fish from below to above average.
- Increase direct landings abroad.

These two objectives might translate into the following targets:

- Increase the landed value of fish from a value which is on average 5% below the average for the fleet as a whole to 5% above the average value in 3 years.
- Increase direct exports from 100 tonnes per year to 150 tonnes per year.

   An example for a fish processing company might be as follows. The company is engaged in traditional curing of herring for domestic consumption, has analysed its internal and external environments and concluded that the penetration of an export market, such as Germany, is practicable. The business has distinctive capabilities, including the ability to process and package cured herring in the required manner. It also has established a good working relationship with an importer. Thus it would have gone ahead with its export trials. Its objective is then to sell some of its product in Germany. Thereafter its target might be to increase its sales of cured herring in Germany to 50 tonnes per year in 3 years.

   An example from the wider economic environment is Marks and Spencer. It was reported to be considering moving into the Japanese clothing market. Although no final decision (*The Financial Times*, 5 November 1997) has been taken, this implies levels of planned capital expenditure (£2.1 billion over 3 years) which can be estimated from the expected increase in the number of outlets and any other implied expansion; thus an objective is turned into a target.

Production

A similar procedure can be applied to the production side of the business. Some typical objectives for a fish farming business might be as follows:

- Reduce the unit cost of salmon produced.
- Improve the quality of salmon produced.

These might then translate into targets as follows:

- Reduce the unit cost of salmon produced from $2000 to $1800 per tonne within 3 years.
- Increase the quality of fish in 'A' category from 10 to 15% of total production within 3 years.

For a fish processor some typical production objectives for the business might be as follows:

- Increase the fillet yield on demersal species.
- Increase the productivity (output per worker) on the canning line.

These might then be translated into targets as follows:

- Increase the fillet yield on demersal species from 45% of the gutted weight of fish to 47% in 2 years.
- Increase the output per worker from x cans per hour to y cans per hour in 3 years.

Institutional and organisational

A business may also have institutional and organisational objectives and targets. Two examples of objectives are as follows:

- Privatise the business.
- Introduce a new, flexible organisational structure.

The corresponding targets might be as follows:

- Offer a fully operational commercial company to selected shareholders within a 2-year time scale.
- Within 2 years have in place a new, non-hierarchical organisational structure.

The business plan

The heart of the business plan is the quantification of the implication of the business strategy. The main purpose of the following example is to illustrate in a very simple format how strategic choices can influence the ultimate measure of business activity – the balance sheet. The balance sheet measures the value of the company to its owners, the shareholders, who are, in business finance, assumed to be interested in increasing that value. Chapter 16 includes a detailed example of the quantification of a business plan and the Appendix to Chapter 7 includes two summaries of examples of business plans prepared for the sector. Birley (1979) includes a very interesting aquaculture example (pp. 36–51); Barrow *et al.* (1992), West (1988) and Williams (1998) have examples of plans.

Simplified example: a divestiture scenario

Suppose that an eastern European fishing company has a strategy for scrapping half its fleet of ten large trawlers. The business plan would predict the outcome of this decision on the company's cash flow – meaning precisely what it says – the impact of the decision on the cash coming into and leaving the company. In this case it will be cash in for the vessels. It would then translate this into the predicted profit and loss account – meaning the cash flow account adjusted for any non-cash changes in asset and liability values. In this case the main non-cash change is the level of depreciation. As some of the fleet will have gone the depreciation as a measure of capital consumption will fall. In other words, the capital assets of the business are costing it less in depreciation because there are fewer of them. The impact of the predicted profit and loss account on the balance sheet is then estimated. The profit (or if there is a loss, the deficit) from the profit and loss account is carried over to the balance to increase (or reduce) the value of the company to its owners. This statement of value is the balance sheet. In this case the company is going to scrap half its aging fleet and replace it with cash. If the valuation of those vessels was correct the effect of this on the balance sheet should be neutral because as one asset goes out another comes in. However, if the vessels were overvalued then the balance sheet may look worse than before because the cash coming in will not compensate for the loss of assets. However, in the longer term if the remaining vessels should be profitable, this profit will feed back into the balance sheet, enhancing the value of the company to its shareholders.

The simplified example below illustrates the case described. Table 3.1 shows that cash flow, at $1 million, is positive in the current year (Year 0). This is the result of fish sales amounting to $5 million whilst operating costs are $4 million. However, depreciation (or the estimated loss of value of the company's assets over time) wipes out the cash surplus; thus the business makes no profit. The balance sheet for the end of Year 0 reflects both the cash inflow of $1 million and the depreciation of the assets. Thus at the end of Year 0 the company's accountants judge the fleet of ten vessels to be worth $10 million. Plus $1 million from the cash inflow the total assets of the business are valued at $11 million. It is assumed that the business has debts of £1 million; thus the value of the company to the shareholders after deducting these debts is $10 million.

However, the next two columns in Table 3.1 show that all is not well. The old

Table 3.1 Financial analysis of a divestiture decision – no change (in US$).

	Year 0 (current)	Year 1 (predicted)	Year 2 (predicted)
Cash flow			
Cash inflow			
Fish sales	5 000 000	4 000 000	4 000 000
Total inflow	5 000 000	4 000 000	4 000 000
Cash outflow			
Operating costs	4 000 000	4 500 000	4 500 000
Total outflow	**4 000 000**	**4 500 000**	**4 500 000**
Net cash flow	**1 000 000**	**–500 000**	**–500 000**
Profit and loss			
Income	5 000 000	4 000 000	4 000 000
Costs	4 000 000	4 500 000	4 500 000
Profit before depreciation	1 000 000	–500 000	–500 000
Depreciation	1 000 000	1 000 000	1 000 000
Net profit	**0**	**–1 500 000**	**–1 500 000**
Balance sheet			
Fixed assets	**10 000 000**	**9 000 000**	**8 000 000**
Cash	1 000 000	500 000	0
Total assets	**11 000 000**	**9 500 000**	**8 000 000**
Debts	1 000 000	1 000 000	1 000 000
Shareholders' equity	10 000 000	8 500 000	7 000 000

vessels are a drain on the company, perhaps because they are expected to be laid up for part of the year, perhaps because they operate with reduced efficiency, or perhaps because access to long-distance fishing grounds has been denied to them. So higher costs and lower revenues mean that there is expected to be a cash outflow of $500 000 in each of the two following years. After depreciation this translates into a loss of $1.5 million in Years 1 and 2. The effect on the balance sheet is serious. The cash resources are depleted first to $500 000 after $500 000 has been used to meet the cash outflow. Then the same thing happens in Year 2 when a further $500 000 leaves the business. On top of this the assets are depreciating all the time, leaving the value remaining to shareholders in headlong decline with no prospect of improvement. Something must be done. Table 3.2 shows what is predicted to happen if the company scraps its old vessels. Year 0 is the same as in Table 3.1. By the end of Year 1, five vessels are scrapped and the company receives $200 000 for them. However, it has to pay out $500 000 in severance pay to the redundant workers. But the significant saving in operating costs is expected to result in a positive cash flow of $1.2 million. Depreciation is reduced because the number of vessels falls, leaving a net profit of $500 000. The balance sheet in still slightly worse than under Scenario 1 (Table 3.1 – 'no change') because the company is now smaller – five rather than ten vessels – so it has fewer assets. But by Year 2 recovery starts and the residual value belonging to shareholders increases.

Table 3.2 The business plan after divestiture (in US$).

	Year 0 (current)	Year 1 (predicted)	Year 2 (predicted)
Cash flow			
Cash inflow			
Fish sales	5 000 000	3 500 000	3 500 000
Scrapped vessels	0	200 000	0
Total inflow	**5 000 000**	**3 700 000**	**3 500 000**
Cash outflow			
Operating costs	4 000 000	2 000 000	2 000 000
Severance pay	0	500 000	0
Total outflow	**4 000 000**	**2 500 000**	**2 000 000**
Net cash flow	**1 000 000**	**1 200 000**	**1 500 000**
Profit and loss			
Income	5 000 000	3 700 000	3 500 000
Costs	4 000 000	2 500 000	2 000 000
Profit before depreciation	1 000 000	1 200 000	1 500 000
Depreciation	1 000 000	700 000	700 000
Net profit	**0**	**500 000**	**800 000**
Balance sheet			
Fixed assets	10 000 000	7 000 000	6 300 000
Cash	1 000 000	2 200 000	3 700 000
Total assets	**11 000 000**	**9 200 000**	**10 000 000**
Debts	1 000 000	1 000 000	1 000 000
Shareholders' equity	10 000 000	8 200 000	9 000 000

Thus in Year 2 the company keeps its better vessels. They still catch less fish, hence fish sales are $3.5 million and not $4 million as under the first scenario (Table 3.1 – 'no change'). There is a small return for the scrapped vessels, but this is more than offset by severance payments. Even so, by Year 2 the expense of disposals has worked its way through the business and shareholder value is starting to recover.

Economic added value (EAV) comes into this calculation when the profitability of those remaining vessels is examined. The cash flow from these needs to cover all the operating and administration costs, and the depreciation, and any outstanding interest charges. Then the vessels can be said to be adding to shareholder value. It should be added that if the surplus is simply reducing the outstanding debt of the company (that is, the money goes for capital repayments) then that too is increasing shareholder value because it is reducing their debts, which they should be pleased about.

Implementation of strategic decisions

Implementation receives little attention in the business guides, but it is often where the most impressive strategic plans fall apart. Implementation requires very careful planning. It is helpful to think of decision making in business at three different levels. The top level is where strategic decisions are made. The chairman, the managing director and the members of the board of directors are responsible for defining business strategy at this level. They set the rules and policies under which the company operates.

At the second level the strategic decisions of the company are translated into projects and other practical measures. For example, a strategic decision to privatise an enterprise must be planned in detail. Milestones need to be set, resources and responsibilities allocated, progress charts prepared, and monitoring systems put in place. Thus the privatisation process becomes a project.

At the third level various people in the business make day-to-day decisions about their activities. Activities related to a privatisation programme might include the preparation of a memorandum and articles of association for a company that is to be sold to the private sector. For a strategy for increasing sales of cured fish in Germany, examples of related activities might include a particular sales trip or the preparation of a report. This is why *project management* has become a key skill in business today (for example, Buttrick, 1997).

In business, decisions about strategy need to be translated into one or more projects or other practical measures. The process of planning and managing individual projects includes the following basic components.

- A project manager and team must be assigned to the project.
- The project management team must prepare a project plan which is consistent with and contributes to the fulfilment of the company strategy.
- The project manager must arrange for appropriate project costing systems and accounts to be set up.
- The project management team must assign responsibilities for various components of the project plan to different people.
- The project manager must prepare a detailed timetable (usually a bar chart) for the various milestones leading to the completion of the project.
- The project manager needs to have the skills to lead and motivate his project team.
- The project manager needs to set up a monitoring and control system for the project.

For example, in the scenario described, the business objective of a fishing company is to dispose of a number of fishing vessels. The target for disposal is five in 4 years. The various strategic issues, such as the criteria for the selection of vessels to be disposed of, have been agreed. It only remains for the work to be done. The disposal of the vessels can be considered as a project. Someone within the company (the 'project manager') is assigned the responsibility of achieving the target, although in this case the job is unlikely to be a full-time responsibility. He then has overall responsibility for organising and controlling the various activities, such as publicising the offers for sale, perhaps stripping the vessels of useful equipment and gear and organising severance arrangements.

Monitoring and control

Senior management is also responsible for setting up monitoring systems to ensure that projects and other measures contribute to strategic objectives and targets. This can be done by systematic reporting of activities to project managers and then, in turn, the regular reporting by project managers of the successes and failures of the projects. The second strand of monitoring and control concerns accounting data. It is crucial that budgets should be in place for the business as a whole as well as for individual projects and other measures. As the business evolves actual performance must be measured against the budgeted performance and variations from the budgeted performance explained.

Summary

Analysis of the business environment

SWOT analysis – or the examination of strengths and weaknesses (internal) and opportunities and threats (external) – is often used by senior management to examine the interior of the company and the impact on it of the external world.

Objectives and targets

From SWOT analysis a business can proceed to formulate objectives and targets. These should be realistic projections arising out of the in-depth study of the company.

The business plan

The business plan is the quantification of the implication of the business strategy. The main purpose of the plan is to demonstrate how strategic choices can influence the ultimate measure of business activity – the balance sheet. The balance sheet measures the value of the company to its owners, the shareholders. This group of people are assumed to be interested above all in increasing its value.

Implementation of strategic decisions

The implementation of strategic assets should be done by *projectisation* if possible and managed through systematic project management supported by monitoring and control – another responsibility of senior management.

Chapter 4
An Alternative Approach to Strategy Formulation

An alternative approach to strategy formulation has been outlined in Chapter 1. The approach captures the need for fish businesses to think of cooperation and competition (Nalebuff and Brandenburger, 1996), hence its appeal to this author who has seen many examples in the fish industry of where a clear-headed appraisal of the benefits and costs of various strategic options would have made a monumental difference to commercial outcomes. The approach is also a salutary approach because it guards against wish-driven strategic thinking. It forces management to be realistic and honest about what is really achievable.

What does the alternative approach amount to?

Businesses are essentially bundles of capabilities which work better together than apart; the success of a business depends on the alignment of those capabilities with markets; the success – growth and profitability – of a business is constrained by what its special capabilities are. Businesses have to be very careful in deciding which of its capabilities are reproducible, can be purchased readily, and are available to competitors, and which are special, and difficult or impossible to reproduce or copy. Business success depends on exploiting the latter. The best measure of business success is economic added value, or EAV (Kay, 1996).

Distinctive capabilities

The starting point for this way of understanding a business is to identify *distinctive capabilities* deep within the business. The approach is based on the idea that there is never an objectively attractive market for the products of a business for long. Markets are always changing so a business has to examine what it is really good at doing and bring these to an ever-changing marketplace. The idea, then, is that firms should identify their distinctive capabilities, and turn them into competitive advantage through combining them. There are four basic categories of distinctive capabilities: architecture, reputation, innovation and strategic assets.

Architecture

Architecture is the structure of internal and external relationships of a business. Firms buy inputs from suppliers and sell to customers. They also employ people – in a way a kind of supply – the labour input into the business. Part of being a successful firm is to build and sustain networks which discourage defection from the aims of the business. Thus, famous football clubs, such as Manchester United or Liverpool, employ players who are not significantly better than many other Premier League performers, but these clubs get more out of the players than many other clubs will. Consciously or unconsciously these clubs sustain a culture which encourages superhuman performance by the players.

There are well-known companies which get the best out of individuals who, in another context, would probably not perform exceptionally well. Marks and Spencer is an example of a business which induces loyalty from staff, customers and supplies despite being essentially a clothes retailer and grocer. Such companies (Marks and Spencer, IBM, Unilever) avoid abrasive labour disputes and have constructed demanding but productive relationships with suppliers. In the fishing industry the fish selling office with an exceptional engineer establishes loyalty from the fishermen and thus creates success. An exceptional skipper has crew members who stick by him.

Thus architecture makes the whole more than the sum of its parts. It does this through the creation of knowledge and understanding within the relationships of the business. A boatswain and a skipper who know each other well and who know each other's mind will be more effective than a pair who have a merely perfunctory relationship. Factors which might slow down the effectiveness of the vessel are reduced and an ethic of cooperation and working to a common end is created. Organisational routines which contribute to the overall good are established.

In game theory language (see Chapter 5) the Prisoner's Dilemma problem of people placing a priority on their own interests ahead of the collective good is overcome. In addition, the successful businesses, through architecture, create an entity which is very difficult to imitate. The culture is too deeply embedded in the internal and external relationships to be easily transferred. Indeed, the idea that expertise can be bought across company boundaries *and* the business attaching to the person with that expertise is a common error. Only in exceptional cases will this occur.

Reputation

Reputation conveys commercial information to consumers. Reputation is particularly important in service industries where customers only learn about the product through experience. For example, legal services or accounting services are of vital importance to businesses, and buyers of these services do not want to make the wrong choice. Having an auditor with the appropriate name is a prerequisite for potential public companies wishing to float on the stock exchange. Thus, the big name accountancy firms attract fees vastly in excess of the fees small lesser known firms attract because their reputation is worth so much more. In recent years universities, as part of the service sector, have focused on establishing reputations by publishing the results of research assessment exercises and teaching assessments. These are designed to assist potential customers in their choice of tertiary education.

Innovation

Innovation is an essential component of business activity, although it is problematic. Many businesses are based on technological developments and have to update their technological performance all the time. In fishing, a vessel owner who fails to keep up with the latest news on fish-finding gear or the latest in navigation equipment will soon be out of business. However, it is also true that most technological improvement can be copied. The patent system and copyright laws provide some protection for the truly innovative, but copying new techniques with variations is often possible. It is also the case that what may appear to be the rewards of innovation are more properly attributable to architecture because they are the result of a working environment which stimulates the flow of new inventions.

Strategic assets

Strategic assets are different. They do not rely on any intrinsic qualities of the business. Rather they are the name given to those business assets which of themselves enhance profitability. A typical example for a fish farm would be a favourable site with good access to water and an abstraction licence. From the catching sector a typical example is the fishing licence. It has intrinsic value for whoever owns it because it can be turned into a stream of profits with the right management and technology. Other examples include natural monopolies, such as water companies, where the entry costs are so high that it is unrealistic for competitors to emerge. Sometimes businesses own strategic assets because they can supply a label which others cannot. Universities have a strategic asset in that they are entitled to award degrees, which are perceived by prospective employers as a signal of the quality of the person holding one. Bridges and tunnels are strategic assets because the costs of building new ones are high.

From distinctive capabilities to competitive advantage

There are, therefore, four distinctive capabilities: architecture, reputation, innovation and strategic assets. They are very hard to create. If it were easy then the world would have far more Marks and Spencers and Microsofts than it does at present. The firm may formulate an objective in terms of distinctive architecture, reputation or innovation. The question for fish businesses is, therefore, 'What distinctive capabilities do we have, or, if we are really clever, what can we create through the skills and experience that we put together?' Strategic analysis requires the business to look deep into itself and ask some searching questions concerning this topic before it proceeds to develop new objectives. At this point the businessman can go back to his mission statement and ask himself whether he is merely indulging in an exercise in wishful thinking or if he can really deliver good business prospects to his shareholders or himself.

Distinctive capabilities are then brought to the market of choice and become a competitive advantage in comparison with other businesses pursuing the same customers. It is here that the business confronts the marketing problem – actually persuading buyers that the product or service is better, one way or another, than what can be supplied by competitors. The marketing problem is discussed in Chapter 8.

From competitive advantage to economic added value

Competitive advantage becomes EAV when it is translated back into business profits. If it is to make a positive EAV, profits should cover not only running costs and administrative overheads, but also interest rates and depreciation – in economic jargon 'the opportunity cost of capital' plus something more. EAV is the return to the business over and above all these opportunity costs.

Water companies provide an apt example. They incur the costs of collecting and supplying water to industrial and domestic consumers. More than likely they would be the first to admit that in terms of architecture – relationships with employers, suppliers and customers – reputation and innovation they are no Microsoft or IBM. Yet they have proved very profitable and earn significant EAV because they have an overwhelming competitive advantage in the supply of water arising from their monopoly position – the strategic assets of reservoirs and water pipes. Microsoft possesses brilliant architecture. Its relationship with its own employees is excellent, no doubt because many of them have become rich men and women as a result of their choice of employer. Its relationship with its customers, although intensely irritating to competitors, is also excellent, because people find the products of the company attractive and user-friendly. It has a high reputation because product quality is generally very good. Furthermore, it has built up a strategic asset as the main incumbent of the computer software market – it is difficult to sell computers unless they have the Microsoft range of software already installed.

In the catching sector of the fishing industry an important source of competitive advantage, and therefore EAV, is skipper skill. In the industry's company sector the ability of the owner to retain a good skipper is a crucial ingredient of commercial success. In the skipper-owned sub-sector the architecture of the business starts from the bridge where an intelligent and inspiring skipper, and one who is known to fish effectively, has a very good start in attracting the best crew. The success of the vessel depends on the skipper and crew working together – in other words on the architecture of the vessel.

In the fishing industry reputation is usefully applied to some of the service sector. For example, vessel owners frequently need legal advice to help them through some of the legal minefields of domestic and EU legislation. They need good accountants to minimise their tax liabilities and to keep the Inland Revenue away at the lowest possible cost. They need good engineering and boat-building inputs from time to time. They make their choices of the service through the reputation of the suppliers.

Innovation is important for reasons already stated. On the fish processing side a recent topical example is the introduction of electronically controlled machinery to cut and weigh fish fillets into portions with very similar weights. Until recently this was done by hand and the range of outcome in each pack was quite wide. Now, using modern technology to assess the fillet and then cut it, the distribution of outcomes is narrowed down significantly. The problem for the fish processing company is that the buyer, usually a multiple retailer, wants to sell to his custo-mers a pack which is almost identical. Innovation then is turned into competitive advantage and feeds back to the company as EAV. The firm that does not innovate by obtaining the necessary equipment stands to lose market share. However, other companies will follow suit quite quickly and the commercial advantage will soon be lost.

A strategic asset for the fish farm might be access to a particularly useful supply of water or an abstraction licence. A strategic asset for a modern fish-catching company is a fishing licence or an individual transferable quota. These represent sources of potential income beyond any intrinsic ability of the company.

Summary

A complementary approach to strategic planning and management as described in Chapters 2 and 3 is developed by John Kay (1993). The basic idea is that businesses should examine their own unique strengths, use these to develop competitive advantages and turn them into profits. Kay (1993) uses the term 'distinctive capabilities' to describe these special strengths of a business. Dis-tinctive capabilities are:

- Architecture – the structure of networks inside the business and between the business and customers and suppliers.
- Reputation – a fundamental capability for service sector industries like uni-versities and legal firms.
- Innovation – essential but difficult to appropriate in the longer term.

- Strategic assets – the fishing licence, the reservoir or the bridge, which yield value to the company just because they are there.

Distinctive capabilities are brought to the market of choice and, we hope, become a competitive advantage in comparison with other businesses pursuing the same customers. Competitive advantage becomes EAV when it is translated back into business profits. If it is to make a positive EAV profits should cover not only running costs and administrative overheads, but also interest rates and depreciation – in economic jargon 'the opportunity cost of capital'.

APPENDIX: HATIYA ISLAND – A CASE STUDY

This case study is based on a visit to Hatiya Island made by the author as part of a study supported by the Department for International Development in May 1988 (Palfreman, 1988). The abridged version of the study is reproduced here to describe some economic and social aspects of a fishery and to illustrate the application of some of the concepts described in Chapter 4.

The Bangladesh fisheries of the Bay of Bengal

The fisheries of the Bay of Bengal are a *strategic asset* for the surrounding countries, including Bangladesh. They fall into two quite distinct parts. First, a significant fleet of trawlers (about 66) operates from Chittagong, in the deeper waters of the Bay, beyond the 15-m-depth contour. These vessels fish for shrimp and various demersal species. They are subject to licensing, although it is not clear how effective the system of monitoring and surveillance is. If licensing works it enhances the value of the strategic asset. The small-scale vessels are the second major category of vessels, and are estimated by the Department of Fisheries to catch 90 to 95% of the Bangladesh marine catch. The trawler fleet operates well to the south of the range of many of the small-scale fishing vessels, certainly of the non-motorised craft. As the small-scale fleet becomes increasingly motorised, no doubt its range will extend. Many of the small-scale vessels operate in the shallow waters close to the shore of the mainland or of offshore islands.

The fleet of small-scale fishing vessels has developed very rapidly in recent years. The principal development is in the numbers of mechanised craft, 9 to 14 m in length, with inboard diesel engines of between 6 and 45 hp, and there are now well over 3000. *Innovation* is taking place but the benefits will be soon dissipated because vessel owners will copy each other. The Department of Fisheries reports that the size of the non-motorised fleet of about 15 000 vessels, powered by oars and sails, is fairly static.

The principal fishing method used by the mechanised vessels and the larger non-mechanised craft is the drift gill net (*chandi jal*), for use in the hilsa fishery. The hilsa gill net is typically about 1800 m long, with a mesh of 7 to 14 cm (3 to 5.5 inches), although lengths vary considerably, between 800 and 3100 m. Set-bag nets (*behundi jal*) are also common, and are used outside the hilsa fishing season. The *behundi* net is a fixed tapering net, resembling a trawl net, set in a tidal stream. One fishing vessel can operate from two to five behundi nets. These usually have fine meshes and are designed to trap virtually everything that swims into them. Various other nets and gears, such as long lines, cast nets, beach seines and trammel nets, are also used.

The dominant species caught is the hilsa (*Tenualosa ilisha*), comprising about 75% of the catch of the small-scale fishing fleet. Other species of some importance generally include the poa or jewfish (Sciaenidae) and the katanach or catfish (Tachysuridae). Numerous other species are also caught in small quantities.

The Bangladesh hilsa fishery

Morphometric studies show that there may be four morphological groups or stocks comprising the Bangladesh hilsa fishery, but electrophoretic studies distinguished only two types: the riverine and the marine. Information about the spawning and nursery grounds also supports the electrophoretic study. A large spawning ground has been identified between Hatiya, Sandwip and Monpura Islands. The marine stock may migrate from the sea to the freshwater against the current, to spawn in October when the water is nearly fresh. Two large nursery grounds have been identified (Md. Jalilur Rahman, pers. comm.). Estimates of the total Bangladesh annual hilsa catch, from rivers and the sea, vary, but the catch is probably between 200 000 and 350 000 tonnes. Wherever the correct figure lies, it is clear that hilsa may account for about 25% of the total fish catch of Bangladesh. At the time of the study there was no unambiguous evidence of biological over-exploitation of the resource. Concern is expressed about the catching of juvenile hilsa with very fine meshes, the use of bamboo fencing across the rivers and fixed nets in the rivers of Bangladesh (Md. Jalilur Rahman, pers. comm.). Anecdotal evidence certainly suggests economic overfishing, even if not biological over-fishing, because fishing effort has evidently increased without an increase in total fish catch, implying that more fishing vessels than are strictly necessary are operating in the Bay.

Whatever the state of the biological resource, economic overexploitation is predictable and inevitable, unless fishing effort is controlled. The recorded number of mechanised fishing vessels is increasing rapidly and at the time of the study there was observable evidence from the boatyards in Chittagong that this is continuing. As more vessels are mechanised, fishing effort will increase. Against the background of a resource yielding a constant total catch, this implies that returns to additional investment will decline, and the overall profitability of the fleet will fall. The net result will be that the fishery will operate at a higher cost than necessary, and the fishermen themselves will be impoverished further as costs rise and productivity falls.

Hatiya Island

Hatiya Island lies in the Bay of Bengal in the south of the district of Noakhali a few miles from the mainland. The surrounding water is very shallow, all within the 10-m-depth contour, and is not thought of by local fishermen as 'sea'. Rather it is 'river', and, for the part of the year when the rivers are in flood, is brackish rather than saline. Hatiya Island and other low-lying islands at the mouth of the Ganges River system are accretions of alluvial deposits that appear and disappear over quite short periods of time. Hatiya itself is only 300 years old and is changing shape quite rapidly. The north of the island is losing land each year, but at the same time the island is gaining land to the south and south east. Nizum Dwip is a recently emerged island to the south of Hatiya. Land erosion is often quoted as the principal economic problem of the island. The physical appearance of Hatiya is like many other parts of Bangladesh. The land is flat and is almost entirely used for the production of paddy. Some of the land enclosures are quite large, while others are small. The landscape is broken by clumps of trees and coconut palms. There are more trees in the older, north of the island than in the south. The northern coastal margins are very low, with visibly eroding cliffs, whereas in the south the cultivated land gives way to some tree plantations (by the Department of Forestry) and mangrove swamp.

The island has an area of over 1000 km^2 (nearly 400 miles^2). At the time of the study it had only 4.8 km (3 miles) of metalled road and about 16 km (10 miles) of brick-sealed road. The many remaining roads and paths have baked clay surfaces, which rapidly become unusable in wet weather. Limited transport to the mainland implies that Hatiya is quite isolated. The population (300 000 at the time of the study) relies on local resources for the bulk of its needs. Agriculture is the main industry. There is some fishing, trading and small services. Hatiya is a food-surplus area. Paddy, poultry, groundnuts, oilseeds, cattle, goats and fish are all sent to the mainland for sale. There is also some migration from the island, as people leave to find work on the mainland.

Most of the land on Hatiya is 'controlled' by a small number of families. Although land has been subdivided through inheritance, the 'wakfah' system, which limits the persons to whom land may legally be sold, ensures that land may not be sold outside the family. Family members work closely together. Ownership and control of land is the basis of their economic and social power. Other activities, such as the ownership of fishing vessels and fish-carrier vessels, shops or other trades, are sidelines. The society on Hatiya is predominantly Muslim, with an estimated 20% Hindu minority. Social attitudes are, in general, conservative and traditional. There are about 50 policemen on the island. The small number of policemen relative to the population was quoted as evidence of the law-abiding character of Hatiya people.

The fishing industry of Hatiya

The fleet

There were no accurate estimates of the size of Hatiya's fishing fleet at the time of the study. An informed guess put the number of larger, non-motorised craft, varying in overall length from about 12 to 15 m (40 to 50 ft), that is those suitable for the hilsa fishery, at around 500. During the hilsa season, May to November, they are rented by fishermen. In addition to the 500 larger non-motorised craft, there may be about another 500 smaller vessels, also mainly country craft of a traditional design. These do not go fishing for hilsa but operate the set-bag net (*behundi jal*) for most of the year. It seems that ownership of the smaller vessels is more widely spread and many fishermen said that they owned their own vessels. There are also an unspecified number of mechanised vessels, some of which are used for fishing, especially hilsa, whilst others are used as fish carriers. One estimate put the number of mechanised craft on Hatiya (fishing vessels, carrier vessels and others) at 100 to 150.

Fishing methods and nets

Hilsa nets (chandi jal) are large. The Hatiya fishermen interviewed for the study said that their nets were around 1000 m and others have suggested a larger figure of up to 3000 m. A net would typically be obtained from Chandpur or Dhaka. At the time of the study a complete net would cost from tk (taka) 80 000 to tk 100 000 (about $14 000). Some of the fishermen said that the advance from the *mahajon* (the vessel owner) is often used to purchase a net; others said that they would hire a net for a season.

Fishing for hilsa takes place within 24 km (15 miles) of the shore of Hatiya, in water depths of up to 10 m. Sometimes it will take place only a few yards from the shore. Vessels go to sea whenever possible, and fishermen said that during the hilsa season bad weather would normally only prevent fishing for 15 to 20 days in a year. Estimates of catches by weight are difficult to establish because the fishermen measure the catch by individual fish, but one estimate was that the annual catch by a vessel might be about 12 or 13 tonnes.

The earnings of the vessels are divided into two equal shares, one for the vessel owner and the other for the crew of 10 to 12 men. After the repayment of advances, the crew share is divided equally between crew members apart from the skipper (*maji*). He receives more than other crew members; a figure of 50% more than the share of a crew member was mentioned by one fishermen's group.

Outside the hilsa season the predominant fishing activity involves the use of the *behundi jal* or set-bag net. Many hilsa fishermen turn to this type of fishing during the winter months. There are, in addition to the hilsa fishermen, many Hatiya fishermen for whom the operation of the *behundi* net is the main fishing activity.

The smaller non-motorised craft carrying crews of about seven generally use this fishing method. One vessel takes about five nets which are then set at favourable points. Fishermen interviewed also referred to a larger mesh drift net, the *koral jal*, used in the winter from November to January. This is 150 to 210 m (500 to 700 ft) long and is used for catching catfish.

Fish selling

Fish selling at first hand usually takes place on the water. There are 10 to 12 locations around Hatiya where the fishing vessels may go to sell their fish. When the vessel is ready for a sale the skipper hoists a flag, the fish-carrier vessel approaches and the deal is struck. This is often a competitive sale. Fishermen noted that as many as five to ten mechanised fish-carrier vessels might offer to purchase fish from a fishing vessel, and the skipper is often in a position to bargain to get a better price. Fish is passed by crews from one vessel to another in baskets. Occasions arise when no carriers are available, perhaps because of a glut or because the vessels are occupied elsewhere. These cases are quite infrequent, according to the fishermen interviewed. If no sale is made the hilsa may be cut and salted for later consumption or sale.

It is a rule of the river that when there is a financial arrangement between a fishing vessel crew and the owner of a mechanised fish carrier, the fishing vessel must sell its catch to the carrier, if he is available when the fishing vessel is ready to sell. The price discount the carrier receives is effectively the interest charge on the loan taken out by the fishing vessel at the beginning of the season. This obligation to sell to the fish carrier lasts as long as the loan is outstanding, but when the loan is paid off it lapses. The discount varies. Examples quoted by fishermen were that if the price per pawn (80 fish) is tk 600, the vessel would expect to receive tk 500, but if the going rate is tk 1200 per pawn, the price received would fall to tk 1000. From the point of view of the fishermen the advantage of the arrangement is that they obtain working capital for vessel and net hire at the beginning of the hilsa season. From the point of view of the fish carrier/owner, entering into this kind of arrangement is a method of securing supplies. Those fish carriers with the security of supplies have a substantial competitive advantage over those without.

The fish carriers transport the fish to one of the main markets to the north. The vessels sell the fish and purchase ice, before returning to the fishing grounds. It is almost certainly a fallacy to assume that fish transportation from the fishing grounds to the major markets is always profitable. It is a competitive business and involves risk. One mechanised vessel owner the author met had opted out of fish carrying and said he preferred to use his vessel for fishing, which was more profitable. The fish carriers face risks in at least three respects (in addition to any financial risk incurred in giving loans at the beginning of the season):

(1)   There is the risk, already noted, that the vessels will cruise around, using fuel, incurring labour costs, etc., trying to purchase fish from vessels already committed to other buyers or not fishing successfully.

(2)   The carriers face the danger of the deterioration of fish quality. If fish is already of uncertain quality when purchased from the fishing vessel it may be in an unsaleable state by the time it reaches the major markets. Fishermen confirmed that this could occur if there was a delay in the arrival of the carrier vessel, if the carrier was carrying insufficient ice, or if the fish was badly stowed so that some of it became warm.

(3)   The carriers face the risk of a poor market. The carrier skipper must make his choice about where to sell the fish and hope that prices are high enough to meet his costs. In a fish glut (at times of the *joo*) prices at the markets can fall to very low levels.

Infrastructure and marketable inputs

There are no jetties or quays on Hatiya. Vessels are protected by being dragged onto mud banks or the dry land. Moreover, the instability of the land would probably render futile any attempt to provide more permanent structures. No ice is manufactured on Hatiya and there is no cold storage or chilling facility. Local ice manufacture may be feasible, but there would be difficulties in transporting ice to vessels in the absence of all-weather roads or other means of transport. The method of obtaining ice – purchase by the carrier vessels in the main markets – may be more efficient than any alternative.

Dwip Unnayon Sangstha (DUS)

Dwip Unnayon Sangstha means Island Development Association. It is a non-governmental organisation (NGO) which was evidently well led, efficient and honest. At the time of the study it was almost entirely funded by Oxfam who showed great confidence in it.

Organisation of DUS

DUS is an officially recognised NGO, and, as is required under Bangladeshi law, subject to routine annual audit. It has a 'General Body' consisting of about 15 members who are leading local figures, including a school teacher, a social worker, businessmen and landowners. According to the rules of DUS, the General Body must meet at least once a year. DUS has an 'Executive Committee' appointed by the General Body, meeting at least three times a year.

The organisation has a pyramid structure. At the base of the pyramid are the target groups. These consist of village people who have agreed to come together for the purpose of pooling savings, engaging in group activities and borrowing for small projects. At the time of the study there were 246 target groups, with an average size of 15. Of these there were 47 fishermen's groups. The main goal of DUS is to make these groups self-reliant. After a few years it hopes that groups will depend less on loans from DUS for income generation, and that the money required for small investment projects will come entirely from their own resources.

The Project Coordinator is the secretary of the organisation and runs its administration. He is responsible for head office activities, such as training courses, accounting, the supervision of project monitoring and support services. Legal advice is especially helpful for the extreme poor in cases of disputes over land and property. The Executive Director is the political leader of DUS. He is responsible for relations with donors, government, local government and others. He finds the resources for the NGO to operate and decides on appointments.

Philosophy of DUS

The ideas of the Brazilian educationalist and philosopher Paolo Friere are the inspiration behind DUS, and many other NGOs in Bangladesh. The word 'conscientisation' has been coined to describe the process, inspired by Paolo Friere's ideas, by which men and women achieve a deepening awareness of the socio-economic reality which shapes their lives. Through understanding and acting upon the problems that face them they become aware of their capacity to transform reality and achieve greater self-reliance.

This approach translates into the following statement of objectives prepared by DUS:

- to organise target people into groups for collective action and to raise their socio-economic awareness;
- to encourage groups to raise their own funds through a weekly savings programme;
- to encourage groups to meet regularly;
- to provide innovatory loans to assist groups in projects to raise family incomes;
- to run training programmes for target groups with the dual aim of developing skills as well as strengthening socio-economic awareness;
- to run functional educational programmes to eradicate illiteracy;
- to encourage target groups to perceive underutilised local resources and to make them able to plan small-scale projects of the appropriate size for the utilisation of such resources;
- to create employment opportunities for rural people;
- to encourage the participation of women in national development;
- to improve family health care, health education and family planning.

Commentary on DUS management practices

Loan recovery

DUS divides Hatiya Island into a North Zone and a South Zone for administrative purposes. Any outside observer would have been concerned by the low rate of recovery of loans in the South Zone (46%). Even the figure of 90% in the North Zone was inadequate. In the context of the low rate of recovery of loans in the South Zone the idea of a 'revolving fund' is nonsense. At the time of the study DUS said that it had taken steps to rectify this, by locating an office in the south of the island.

Interest rates

DUS levied a service charge of 12% on its income-generating projects. Four percent was paid into the DUS revolving fund, 4% to a social fund, and 4% to a suspense account to be repaid to the group when the loan is paid off. So only 4% of the charge went back into the NGO's capital, whereas it should have been getting at least 12% (the rate at which money was losing value through inflation) if the value of its capital assets was to be maintained. At the time of the study the rate of inflation in Bangladesh was around 12%, so money was losing that amount of value every year. An organisation which wishes to conserve its capital and make a contribution towards its costs should charge more than 12%.

Banking policy

Loan recovery and interest rate policy suggest that DUS is operating with goodwill towards the poor, but without a clear picture of its commercial policies. If it is the intention to subsidise the poor, then a consistent policy of interest rate subsidies would be one way to do it. However, it must then be made clear that it is dissipating its capital. If it is the intention to operate as a bank, in a commercial or quasi-commercial manner, then DUS should charge a rate of interest that reflects the need to conserve its capital, to meet any costs properly attributable to the banking function, and to reflect the risks involved.

Strategy of DUS

Architecture

DUS intended to strengthen its relationships with its actual and potential fishermen-members by providing some of them, on favourable terms, with a low cost unit of capital – a fishing vessel. A donor would provide the funding, DUS would organise the construction of the vessels on Hatiya, and it would sell them to fishermen, and use the resulting funds for further development.

DUS also intended to strengthen its relationships through the provision of a fish-carrier service – effectively subsidised because a donor would provide the vessels. Would fishermen sell to the DUS carriers? Quite probably many of them would, provided they could see a long-term gain from doing so. However, DUS would have to provide an efficient service.

DUS would be entering a very competitive relationship with the current fish distributors. Herein must lie risks which would need to be considered seriously in any strategic plan. Even with a capital subsidy, was it really plausible that DUS, essentially a social welfare organisation, would be able to withstand the full onslaught of serious games of 'chicken' with some of the potential commercial rivals?

Reputation

DUS is a service organisation whose long-term income depends on a good reputation with its members and external supporting organisations. It might well lose its very good reputation for improving the social welfare of the extreme poor on Hatiya if it becomes distracted by attempting to make a commercial success of its fish business.

Innovation

DUS has not overtly considered innovation, but it should. If it can find innovative methods for catching or marketing fish it might find itself able to compete more effectively against the existing trade. For example, if it could find new markets for hilsa in India, which yielded a better return in exchange for higher quality handling and market organisation, it might strengthen its hand. However, if it does this then it is also obvious that its innovations would be easily copied by others.

Strategic assets

In the long term DUS should be aiming to establish itself with strategic assets. It might do this by supporting and then becoming a key player in the issuing of fishing licences in the Bay. At the time of the study it was already becoming active in land issues, securing places on the relevant committees. Sooner or later fisheries management must come to the fisheries of the Bay and DUS should aim to be a centre of developments, thus securing control over strategic assets for itself and its members.

Strategy of the fishermen

Architecture and reputation

The fishermen of Hatiya have a small competitive advantage because they know how to fish, and potential suppliers of fishing vessels know that they know how to fish. Thus the structure of relationships supplies them with a competitive advantage and therefore added value. It is relevant to ask whether anything can be done to strengthen their relationships with fish buyers. In other fisheries, individual skippers can improve their returns by handling fish better, or entering into longer-term contractual relationships with buyers.

Innovation

One of the major issues for the fishermen will be whether to mechanise or not. If there is any current uncertainty in the fishing community about this (as there certainly is on the west coast of peninsular Malaysia), then this is precisely where DUS can make representations to secure permanent protection for non-motorised vessels.

Strategic assets

As far as fishing activities are concerned, the fishermen play a Prisoner's Dilemma game (see Chapter 5). They have no choice but to compete for supplies, because if they do not fish as well as they can, others will do so. In this respect allegiance to DUS may help them to change the rules of the game so that they can begin to develop a fishery that is a strategic asset, to be nurtured to provide a better long-term income.

Chapter 5
Cooperation or Competition?

Chapters 2 and 3 introduced readers to strategic planning and Chapter 4 developed an alternative approach to strategic thinking. This chapter develops that theme and offers readers a special angle on strategy which the author found useful, for reasons which will become clear.

To compete or engage in collective action is a major strategic issue for businesses and individuals. The question was first explored by Olson (1965). One of the most difficult strategic issues faced by fishing vessel operators, and, in different contexts, fish farmers, is whether to compete or cooperate. A dilemma facing people in the fisheries sector all over the world is whether to cooperate in collective institutions or to compete with one another. Fortunately economics offers a conceptual framework which can help businessmen in fisheries to decide – this is called game theory.

The most acute problem for fishermen is often the question of resource management. It is easy for administrators to blame the industry for the state of fish resources, but there may be something the catchers themselves can do about it, in certain circumstances. The use of game theory to study this has become well known, particularly through the work of Ostrom (1990) who shows how different game structures can produce different outcomes in the management of common property resources.

Game theory

Game theory is a branch of applied mathematics which studies how people and businesses make decisions when they are mutually interdependent. This means that the player does not know for definite what the outcome of his strategy in the game is going to be, because that depends on someone else's strategic choice, as well as his own. Indeed, a better name for game theory would be the *theory of interdependent decision-making*. For readers looking for rigorous and relevant economic expositions Cornes and Sandler (1996) and Sandler (1992 and 1997) are recommended.

The relevance of game theory to strategic choices by business has only recently taken hold in business analysis, pushed along most entertainingly in Kay (1993) and Nalebuff and Brandenburger (1996). This is partly because game theory is often expressed in the language of mathematics or more abstract economics. To illustrate, consider the following statement, selected from numerous others in an introductory textbook on the topic:

> 'A trembling hand perfect equilibrium is the limit of the sequence of Nash equilibria in perturbed versions of the game as trembles go to zero' (Hargreaves Heap and Varoufakis, 1995, p. 69).

What this really means is that one of the ways of generating a mathematical solution to a game theory problem is to introduce the concept of the 'trembling hand' – essentially an error, or lapse in motivation – in the strategies of the players so that they do not always behave in a predictable way.

From the point of view of real-life strategic choice the idea of the odd deviation from what is expected is an interesting train of thought. An example might be the fishing skipper who normally sets his course and proceeds to the fishing ground of his choice. Others follow because they recognise his talents. So occasionally he does something different and confuses his rivals, and the game between the competing fishermen has an unexpected outcome. However, with the greatest respect to fishing skippers, would the above quotation from Hargreaves Heap and Varoufakis have made much sense to them?

The fish industry abounds in situations where the choice for individual businesses is either to compete or collaborate, or a mixture of the two. Some examples follow.

- Fishermen invariably supply mutual assistance at sea (or other water body) when lives are in danger. This is true between even the most bitter rivals.
- Many fishermen belong to producers' organisations, such as EU-style Fish Producers' Organisations, to implement common rules of supply.
- Fishermen race back to port, to try to be first on the market. However, they cannot leave the grounds too quickly before they have a reasonable catch. Thus, they compete to get back quickly and collaborate in their acceptance of the common selling rules.
- Fish farmers join forces in national associations to improve their marketing.

Basic models

A review of the literature on game theory reveals the vast number of different interactions that have now been modelled by the specialists. However, there are certain basic structures which are generally sufficient to analyse most social and economic interactions in practice.

The Prisoner's Dilemma

The most commonly quoted is the *Prisoner's Dilemma* (or PD). In this game two criminals guilty of one crime and suspected of another are separated and encouraged to incriminate the other player by the offer of being let off the first crime. Figure 5.1 outlines this game and a number of other different common scenarios.

A rather compelling application of the PD model is the following story, as reproduced in and adapted from Poundstone (1992). Suppose that Emma has stolen a very expensive diamond and is trying to sell it. Through the underworld grapevine she hears of a possible buyer – a well-known Hull criminal, Otto, who is notoriously greedy, duplicitous and shrewd. Otto wants to buy the diamond and Emma wants to sell it and through the underworld intermediaries they do a deal. Otto will pay £500 000 and Emma will relinquish the diamond.

Emma knows that Otto is a liar. She has heard rumours that on many previous occasions Otto has offered money for stolen goods and agreed to hand over the money in a remote place, to guarantee the secrecy of the transaction. But instead of handing over the money Otto has drawn out a gun and killed the crook, walking off with the goods and the money. To avoid this she proposes to Otto that he should leave the money in an attaché case buried in a field to be picked up later, thus avoiding the face-to-face contact that carries so much threat. Similarly, she would leave the diamond hidden in another field for Otto to pick up later. When the money and diamond are safely buried Otto will telephone Emma on his mobile for the exchange of relevant information.

Will the arrangement work? It occurs to Emma that she could, if she were so minded, deceive Otto by telling him the diamond is buried but actually hanging on to it. Thus she would get Otto's money as well as keeping the diamond. A similar argument crosses Otto's mind. He could lie to Emma, tell her that the money is buried somewhere, but actually keep his money and the diamond. Emma and Otto are both intelligent people and appreciate that the other might deceive them. Reaching agreement that they will does not really help to resolve matters because there is always the possibility at the back of each of their minds that the other is not strictly honest. The upshot is that the transaction does not take place. Otto and Emma would have gained by the simple straight transaction – Otto wants the diamond more than the money and Emma wants the money more than the diamond – but mutual mistrust prevents the transaction taking place. It becomes clear to the observer that the problem is that there is no institutional setting under which the transaction could take place and would then be policed to ensure adherence to its terms.

A change in the pay-off structure alters the game. If the payoff structure is altered such that Emma is too frightened to deceive Otto – she fears the long-term consequences of her deception and, similarly, Otto is wary of breaching the agreement – then agreement might be reached. This game is known as 'The Criminal's Revenge'.

The applicability of the Prisoner's Dilemma game to some problems is clear. Consider the case of a voluntary minimum price scheme run by a Fish Producers' Organisation (FPO). Basically, if fish fails to attract a buyer at the minimum price

Deadlock

	Otto				From Emma's point of view
	Cooperate		Defect		
Emma	Cooperate	0	0	−1	3
	Defect	3	−1	2	2

DC>DD>CC>CD (no incentive to cooperate)

Prisoner's Dilemma

	Otto			
	Cooperate		Defect	
Emma	Cooperate	2	2	−2	3
	Defect	3	−2	−1	−1

DC>CC>DD>CD (defection dominant)

Chicken (Hawk/Dove, defect = hawk)

	Otto			
	Cooperate		Defect	
Emma	Cooperate	1	1	2	0
	Defect	2	0	−1	−1

DC>CC>CD>DD (defect and Emma wins, as long as Otto cooperates)

Stag Hunt

	Otto			
	Cooperate		Defect	
Emma	Cooperate	3	3	0	2
	Defect	2	0	1	1

CC>DC>DD>CD (an assurance game, provided there is mutual trust)

Battle of the Sexes

	Otto			
	Cooperate		Defect	
Emma	Cooperate	1	1	3	2
	Defect	2	3	0	0

CD>DC>CC>DD (coordination is the problem here)

Leader

	Otto			
	Cooperate		Defect	
Emma	Cooperate	1	1	2	3
	Defect	3	2	0	0

DC>CD>CC>DD (also a coordination problem)

Fig. 5.1 Game theory outlines.

then it is withdrawn from sale by the FPO and the supplier receives no compensation. Suppose there are just two members of the FPO. The members of the fish producers' organisation gain collectively if they adhere to the minimum price, but in the event of fish failing to attract a buyer the fisherman is invariably tempted to defect and get the best price he can, even if it is below the minimum price.

The problem of fisheries management is also susceptible to analysis using game theory, though it is not always a PD problem, but this example is. Consider the case of a fishery under which any addition to fishing effort reduces catch per unit of fishing effort (that is, for example, one trawler catches 100 tonnes per year, the next one catches 90 tonnes per year and the catch of the first is reduced to 90 tonnes per year). So long as each new vessel entering the fishery adds to the total net economic benefit from it there is a collective gain from additions to fishing effort. However, beyond a certain point new vessels entering the fishery start to reduce the total surplus although they might be individually profitable. This is where the PD analysis is applicable. The best outcome for the individual is that everyone else in the fishery should restrict their effort whilst the individual (Otto!) operates flat out. The second best outcome is some form of cooperative solution. The third best is mutual defection because then at least no-one is taken for a ride. The worst outcome for an individual is that he should restrain his fishing effort whilst no-one else in the fishery does, because the outcome for resources is the same, but he also suffers from relative deprivation.

The Prisoner's Dilemma applied to large numbers

The extension of the PD model to large numbers (the *n*-person game) is a simple extension of the two-player model (Sandler, 1992, 1997; Cornes and Sandler, 1996). Consider the case of a fishery with several fishing companies. The fishery would operate more efficiently if effort were less and, to the extent that any representative individual company reduces effort, the others all benefit from higher catch rates. How will the representative company respond? The dominant strategy of each company is to carry on fishing at the original level of effort, hoping to free ride on the effort-curtailment of one or more of the others. This corresponds to the behaviour of Otto and Emma. The upshot is obvious. No company is going to reduce fishing effort voluntarily (Cunningham *et al.*, 1985).

The criminal's revenge

The Criminal's Revenge is the result of a change in the pay-off structure such that defectors are penalised for defection. It is called 'The Criminal's Revenge' because it presumes that those who fail to stick by the code of practice of criminals that they will not incriminate their partners in crime are penalised. The Mafia is well known for the imposition of this type of incentive to cooperate. In a fisheries situation the severe financial penalties that have been applied to fishing vessel owners in breach of regulations are an excellent example. The structure of the game is the same as 'Stag Hunt', outlined below, because there is a strong incentive to cooperate.

Chicken

The basis of the game of chicken is the case of the game in which two fool-hardy (or thrill-seeking) drivers point their cars at each other and drive fast. There are four possible outcomes to the game: (1) head-on collision; (2) both drivers swerve; (3) one driver swerves and the other proceeds; and (4) one driver proceeds and the other swerves. The best outcome from a driver's point of view is that the other driver should swerve. Thus, he not only stays in one piece, but also demonstrates conclusively that he is not a 'chicken' (coward). The worst outcome is that both drivers fail to swerve, because this will result in a collision. The intermediate outcome is that both swerve. A driver can strengthen his hand by ripping the steering wheel off and ostentatiously hurling it out of the window of his car.

Followers of international diplomacy will be well aware of the game of chicken. The Iraq crisis of February 1998 is a classic example. The international community was faced with a threat by the government of Iraq to develop certain types of weapons which are widely perceived as being a threat to the security of the rest of the world. The rest of the world, to a greater or lesser extent, led by the United States, threatened Iraq with a bombing campaign. The United States and the UK wished to see the agreement between the Secretary General of the UN (Kofi Annan) and the government of Iraq as a swerve by Iraq. The response to the agreement clearly suggests this, although both governments also appear to recognise that Iraq's swerve leaves open the possibility of more plays of the game.

The most celebrated example of an international game of chicken is the Cuban Missile Crisis of 1962. The United States and the former USSR (Union of Soviet Socialist Republics) came closer on this occasion to nuclear war than at any other time, before or since. Cuba had been a growing irritant to the United States since 1959 when Fidel Castro came to power through a revolution. In the summer of 1962 American surveillance aircraft observed nuclear missile bases under construction in Cuba. They were being constructed by the USSR and would have resulted in missile sites a mere 145 km (90 miles) from Florida. On 22 October President Kennedy acted. Knowing that more Soviet ships were approaching Cuba, Kennedy announced a naval blockade of the island. The announcement generated an equally intransigent response from Nikita Khrushchev, the leader of the Soviet Union at that time, who insisted on going ahead with the construction and threatened to use force to achieve his objectives. The rest is history. On 28 October the Soviet government offered to dismantle the sites, getting nothing in return.

A business example of the game of chicken is quoted by Kay (1993, p. 46). In 1986 deregulation of the London financial markets allowed any business to buy and sell government bonds on its own account – in the jargon, to become 'market makers'. Twenty-eight firms, including many major banks, decided to do so. However, after several years many of the businesses had accumulated substantial losses and a number of them had dropped out. They all entered knowing that the industry could not afford to support all of them, but they could also not afford to be left out of the game – to concede territory to rivals.

The Grimsby seiners blockaded the commercial docks in 1978, hoping to induce a swerve by the government in the conduct of its fisheries policy. The threats of legal action against the fishermen were eventually enough to induce a swerve by the fishermen. The cod wars with Iceland were examples of games of chicken, interesting because it was the smaller player who won the wars in the end. Iceland's greater commitment to victory and its international support were sufficient in each case for the British government to swerve and go for an agreement rather than remain and fight the war.

This game is sometimes known as 'hawk/dove'. The government as hawk finally induced dove-like behaviour by the Grimsby seiners. However, over Iceland the British government eventually became a dove to the Icelanders' hawk. Sometimes this makes sense when the alternative is mutual destruction.

A study of fishermen on the west coast of peninsular Malaysia

A study of fishermen on the west coast of peninsular Malaysia (Kamaruzaman, 1997) reveals just how complex the institutions governing fishermen's behaviour can be, especially once one departs from the analytically comfortable open-access scenario. Where rules are established and the Malaysian fishermen know the reasons for them, cooperation is the normal pattern of behaviour. However in one significant respect – engine size – there were no institutional rules in place and the pattern of behaviour corresponded most closely to the 'chicken' game. Some had invested in larger engines, and this constituted a kind of dare to those with smaller engines. They wanted to keep using large engines, but hoped that the rest would stick to small engines – so they were playing a kind of game of chicken with the small engine men. For the time being those with smaller engines knew that replacement might damage the fish resources so they were wary of doing so. The 'stickers' were driving hard to advance their technology, but the 'swervers' felt, at least at the time of the study, that they would be better off by keeping to their relatively small engines. It is important to note that the circumstances of the fishery were dynamic and more and more of the swervers were switching sides. In other words, the problem resembled increasingly a PD problem rather than a chicken game.

This study found that the level of awareness about resource exploitation problems among the trawl skippers in the study area was high. The majority of them knew the benefits that they could achieve from collective resource conservation. They also understood the problem of overfishing and the factors causing this problem. A high awareness, coupled with high dependency of their livelihoods on the fishery, made them willing to work collectively to contribute to collective goods, that is, to reduce total fishing effort in the fishery. This willingness is common among fisheries of the world but, as became evident in the study, an institutional setting is needed to bring it to life. In the absence of an institutional setting, the result is mayhem.

The institutional arrangements in Malaysia provided the means to coordinate the interdependent choices of individual skippers in some respects but not others. This can be seen in three collective action situations in the trawl fishery, namely (1) maintenance of a lower number of monthly fishing days, (2) compliance with the time regulation, and (3) compliance with the vessel replacement procedure. In this game, individual skippers have a dominant strategy, that is, to participate in individual fishing effort reduction. Once the regulations on these were in place and accepted, fishermen were content to abide by them.

Stag hunt: an assurance game

The basis of the stag hunt game is that there are evident gains from mutual cooperation. If people work together on the hunt they will catch the stag – the biggest prize. However, if they are diverted from this, perhaps because one of them sees a hare and tries to catch this instead, the prize will be lost. For the player who continues to hunt the stag whilst the other player chases the hare there are considerable costs. Not only does he waste resources on a fruitless task, but also he fails to receive the compensation of the hare. Essentially what is needed is assurance that both parties will maintain the effort to go for the main prize. The game differs from the Prisoner's Dilemma because the cooperative outcome, rather than the cooperate/defect outcome, is the best for each player.

The business world contains many examples of assurance games rather like the stag hunt. It is very important when evaluating whether it is worthwhile joining some strategic alliance to consider whether it is indeed worthwhile sustaining the cooperation for mutual benefit. It often is when there are only a few players. The joint venture between a local company having access to fish resources and an international company wishing to find a profitable use for its vessels is a good example. It will be advantageous for both parties to overcome mutual suspicion and to cooperate for joint gains.

Leader: The problem of coordination

The leader game is best illustrated by the case of two vessels both seeking to enter a harbour at the same time. Each one wants to be first, perhaps so that it can get to the prime site for the fish market. However, if both go at the same time there might be a delay, or a collision. Game theorists introduce the rules of the game to solve problems like this. There might be, for example, an implicit rule that the vessels take it in turns to enter first, or perhaps the vessel owners have agreed a more long-standing approach to the problem. Provided there is some coordination mechanism in place the problem is, in principle, soluble.

Battle of the sexes: another coordination game

A second coordination problem is illustrated by the game known as the Battle of the Sexes. The story behind this game is the picture of the couple, Emma and Otto, who by this time are trying to build up some kind of trust. He thinks that she would like to go to a classical music concert and she thinks that he would rather go to a boxing match. Both of them would rather be together than go to the function of their choice. The result is that Emma decides to go boxing and Otto sits through an evening of Beethoven. So they are both enduring a painful evening – alone. They both have the worst possible outcome because their decisions were not coordinated.

This game has many business applications. For example, the supplier of a consultant's report might think that the client will only be satisfied if he receives a long document. The client, on the other hand, thinks that the consultant will want a free hand to research the subject fully and present the results, but would really rather have a short, punchy, prescriptive report. Thus the consultant labours unnecessarily and the client is dissatisfied.

A common 'battle of the sexes' problem in some offices is the determination of normal working hours for the office. Workers labour late into the night, imagining that this will impress the boss, whilst the boss reluctantly lingers on in his office because the workers are all at their desks.

'Battles of the sexes' often arise when cooperation is needed to achieve a certain objective. In a project, different scientists and technologists need to work together and unless there is some coordinating authority confusion is likely to result. People will arrive on the job when they think they are probably needed and others are called to the project when the project coordinator imagines that they are available.

It is clear that one way of solving this problem is to have someone in charge who is prepared to lay down the law. Thus Emma can insist that they both go to the concert. The client can specify the character of the report required. The manager can lay down the working hours and assure staff that they will not be penalised for leaving at the set time. The project manager has the task of determining the timetable and responsibilities of the project team.

Deadlock

Under Deadlock each player gains if he/she can trick the other party into cooperating, but if both parties cooperate the gains are lost. Thus, Emma hopes that Otto will be foolish enough to cooperate and vice versa. In this game there is not really any incentive for Emma and Otto to cooperate as mutual defection is preferable to mutual cooperation. Two countries competing to control a valuable patch of land (such as an oil field) is an example. Each wants the other to withdraw, but mutual withdrawal from the claimed territory will simply vacate the area with no benefit to either; thus they are deadlocked.

Repeated games

When games are repeated the structure of the problem can alter. There are practical examples of repeated Prisoner's Dilemma problems where it appears that cooperation does evolve. Axelrod, in the seminal work *The Evolution of Cooperation* (1984), was the first to identify this. For example, it seems that German and British soldiers in the First World War evolved behaviour in which the artillery fired over the heads of the infantry, and snipers avoided shooting the medical orderlies who came out into no-man's land to pick up the dead and wounded. The description of the game which approximated most closely to experience and some experimental work is 'tit for tat'. The best rule, in the sense of producing the best outcome for all concerned, turns out to be 'cooperate until someone defects, then defect'.

Consider the case of the withdrawal price scheme, noted as a Prisoner's Dilemma example above. In the long run fishermen gain if people start out by cooperating. If, however, a fish seller forgives defection – someone in breach of the rules – then the rule-breaker will be tempted to do it again. So the fish seller has to break the rules himself if this happens, until the perpetrator mends his ways; then cooperation can start again.

Tit for tat certainly has an intuitive appeal although there is no evidence that this is the ultimate optimal rule in a repeated game situation. This is a subject for evolutionary game theory, which is not pursued in this text.

Making use of game theory

The discussion so far has studied the theory of various games from without. To make practical use of game theory the fish businessman has to look at these various games from within, as a player, and see what can be done to improve his position.

The players

The potential player has to decide whether he is going to join a game. The first thing he must consider is who else is playing. For example, fishermen need to decide whether they are going to enter into some collective arrangement with others in order to press their views on the government or the EU. It will be evident from the games described above that absence of trust is a major obstacle to successful collective action, so a player needs to take action to ensure that his fellow players can be trusted. In all but a very small number of people, such as a family, this is all but impossible without some institutional set-up to enforce it. This is why trade associations and fishermen's cooperatives need to have a constitution – a set of rules which must be followed if people are to participate. Then if one decision goes against the interests of one group or individual, at least the players know that the decision has been reached in accordance with an agreed procedure. A fishing company might be interested in participating in a joint venture. It is well known that trust is important at the outset. However, game theory shows that trust has to be sustained if the mutually cooperative relationship between the partners is to be sustained. Otherwise it becomes in the interests of one or both of the partners to defect.

The reverse might also be true. A fishing company might conclude that it is forever locked in competition with its competitors and therefore the only games it can realistically play, when some degree of interdependence between them emerges, are the Prisoner's Dilemma and the Criminal's Revenge. For example, in an open access fishery with no prospect of management it might conclude that the best solution from a business point of view is to harvest as much as possible in as short a time as possible. In a managed fishery where there is mistrust between the operators it might do well to press for high penalties on the other players for breaches in regulations. The cod wars were played like a game of chicken although in retrospect the UK industry might have achieved better long-term gains if it had adopted a more cooperative stance.

Gains from playing the game

When a company is thinking of joining a cooperative venture with others the upper limit it can expect to get by way of return for playing is the extra value that its participation generates. Its own returns will probably be less than this because the current players will also want some return for cooperating with a new player.

Consider the case of the fisherman who has to decide to join one of two FPOs or remain outside. He has to calculate whether it will be worth his while to join one of them – if his pay-off is going to be worth the membership fee plus any other costs. The members have to decide if they want him in. They will only do this if his presence is likely to add value to their own returns from it.

In a competitive game, such as chicken, the costs and benefits need to be weighed very carefully. For example, many businesses lose a great deal of money because they engage in a prolonged struggle for market share. They need to calculate whether the possible ultimate benefits really outweigh the costs incurred in the battle.

The rules of the game

The analysis of the rules of the game can be simple or difficult. Sometimes it is obvious what is going on. It does not require highly subtle analytical skills to recognise a situation where two companies are locked in battle to drive one of them out of the market, and perhaps out of business altogether. However, in some cooperative games the analysis is more complicated. Again, consider the case of a fishing company making up its mind about joining a producer organisation. A fundamental requirement is that the legal document setting up the institution be scrutinised. It might be the case, for example, that the articles of association provide for voting weighted by value of products landed, or they might provide for equality of voting rights. An agreement between fishermen to divide a quota between them might be analysed as a cooperative game. Whether it is worthwhile for a newcomer to participate depends crucially on the rules by which the quota is shared out and monitored.

The research from Malaysia suggests that the socio-economic situation there most closely resembles the game of chicken. This is not because of any particular aggression but because one group has decided to go ahead with technological improvements – an increase in engine size – and this is seen by the remainder as a possible threat to the fishery as a whole. Moreover, the way the fishery is developing suggests that there is a danger that the situation will evolve into a Prisoner's Dilemma game. The difference at present is that the players would prefer to be in a cooperate/defect situation rather than a defect/defect situation – those who have a preference for smaller engines retain a lingering preference for them – but it will not last for long if their interests remain unprotected. If they have to judge their long-term futures, many feel that it is better if some fishermen remain with small engines. It takes little imagination to see that the government of Malaysia needs to take advantage of this situation and establish the institutional rules to regulate engine size. If they do not then fishermen will come to feel that defection is always better than cooperation and therefore they should also invest in the larger engines.

The players can also sometimes change the rules themselves. A common argument for Britain's membership of the EU is that by being an insider it has a significant influence on the formulation of future rules than if it were on the outside. It is not obvious that this argument holds much water in the case of the Common Fisheries Policy, but it might have more substance in other areas of EU policy, such as the development of common rules on financial markets or the reform of the Common Agricultural Policy.

Creation of collective benefits and costs

In the analysis of games a significant issue is how the benefits and costs are going to be generated. For example, suppose that the government is engaged in a campaign to secure the cooperation of the fishing industry in the implementation of more stringent conservation measures. It is trying to convince the industry that they are engaged in an assurance game, such as Stag Hunt, and it only requires a collective effort for everyone to benefit. It may not be obvious to the representatives of the industry that the proposed measures will actually deliver improved catch rates for them. It is a common error of government departments to present an argument for such measures in terms of stock conservation. However, if there is international competition for the fish or if the expected results of the measures are doubtful or, even, if stock improvement will occur but will not clearly benefit the fishermen themselves, the measures will not be supported.

Business strategies

In a game that is susceptible to game-theoretic analysis businesses should never forget that they can vary their strategies to suit their current advantage. The transformation of assurance or coordination games into a Prisoner's Dilemma is a common occurrence because trust is displaced by mistrust. This breakdown can happen because one of the players decides to cut and run. A minimum price scheme can easily break down because one of the participants decides that it is in his interest to breach the price. In the case of the Malaysian 'chicken' game, fishing vessel owners who are currently acting with caution towards an increase in engine power may legitimately decide that their long-term interests are better served by purchasing new engines themselves.

Scope of the game

An interesting variant on some of the comments given above is that the scope of the game may change or be changed. In the fishing industry an intriguing example of this is the expansion of fish producers' organisations into other countries. Thus fishing vessel owners have extended the scope of their cooperative endeavours (an assurance game) into other countries of the EU because, so they calculate, the management of quotas and withdrawal prices is, in the present context, conducted more effectively across national boundaries. A common way in which producers' organisations have extended the scope of the assurance game that their members play is by entering the fish processing industry in their own right. This extends the scope of the game, thus reaping an additional advantage for their members.

Summary

Prisoner's Dilemma

Defection is the dominant strategy for Emma and Otto because, although there are mutual gains from cooperation, the temptation of defection if the other cooperates is too great. This game represents the genuine paradox.

The criminal's revenge

Defection is no longer the dominant strategy because the consequences for Emma and Otto are harmful, to such an extent that they both prefer to cooperate.

Chicken

Two outcomes are the best for Emma and Otto: one when Emma sticks and Otto swerves and the other when Otto sticks and Emma swerves.

Stag hunt

Emma and Otto both gain from cooperation, provided the players keep their eyes on the ball. This is an assurance game.

Leader

Emma and Otto each say 'me first' but disagree and crash into each other as they meet at a road junction.

Battle of the sexes

Emma and Otto like to go out together, mostly for the benefit of each other's company. Emma goes to the boxing match, really expecting the delights of Otto's company, and Otto goes to the music concert, really to meet Emma.

Deadlock

There is not really any incentive for Emma or Otto to cooperate, as mutual defection is preferable to mutual cooperation. Emma hopes that Otto will be foolish enough to cooperate and vice versa. If Otto does, Emma gains.

Conclusion

A businessman who understands the theory can start to use it. What should he do?

- Understand the personalities and motives of the other players.
- Work out the expected gains from playing the game.
- Ask himself if he can change the rules of the game to his own advantage.
- Ask himself how the benefits of the game are going to be delivered.
- Decide on the appropriate tactics or strategies – which may be mixed.
- Decide on whether the scope of the game is susceptible to change, to his own advantage.

Chapter 6
Setting up a Fish Business

Chapters 2 to 5 have taken a somewhat theoretical look at strategy formulation in the fish industry. This and subsequent chapters move on to some of the more practical commercial aspects of fish business management and development. However, the conceptual framework developed in the early chapters will continue to be useful to illustrate the analysis.

Many small businesses of all kinds fail in their early years. Research (Williams, 1998) suggests that a third of new businesses are no longer in business at the end of the third year, although many of these will be businesses which are sold or are liquidated voluntarily. Fortunately the fish business area is not one where failure is most acute – this is areas such as fashion, transport, light engineering and building. When people are thinking about setting up in the fishing industry they need to consider to what extent they have the distinctive capabilities of architecture, reputation, innovation and strategic assets, explained in Chapter 4. They then need to ask themselves how they compare with others – do they have a competitive advantage? They also need to ask if there is any scope for help – cooperation from others, perhaps on a tit-for-tat understanding or through turning a mutually destructive Prisoner's Dilemma game into something more mutually beneficial (Chapter 5).

What are the options in the fish business?

The catching sector

The fish catching sector is highly problematic for newcomers nowadays, partly because of the technical problems in actually learning how to be a good fisherman. Architecture – input, crew and market management – needs to be learned. Reputation is not so obviously important, but good fish handling practices may help. Innovation is vital because of all the technological improvements in the industry. Also the new entrant must have a licence to fish (a perfect example of a strategic asset), and this might easily cost a great deal more than the vessel itself. The new entrant should also be aware of the various ways in which people in the industry cooperate with each other (as well as compete). An excellent overview of modern thinking about the institutional setting in which fishing takes place is supplied by De Alessi (1988).

The EU is reducing the number of fishing vessels through Multi-Annual Guidance Programmes (MGPs) (see Chapter 11). The fourth MGP has just been announced, hardly an encouraging environment for the new entrant. Clearly, the heavily licensed environment creates *strategic assets* in the sense defined in Chapter 4, so he who has a licence also has a significant head start.

Fish farming

Fish farming is a popular option for many, but it is important to have technical knowledge of the business before plunging in (Bjørndal, 1990). However, as with all other businesses, it is the extra capabilities that the businessman can bring to bear which are going to make the difference. The evidence is growing that cooperation (see Chapter 5) is a fundamental source of commercial gain for small fish farmers. They can work together on markets, technology and input prices to their advantage. Fish farmers face much hard work in a highly competitive environment. They also have environmental constraints and regulations to contend with as well as the risk of disease and the loss of stock. Location, access to clean water and abstraction licences are the strategic assets a fish farm needs.

Primary fish processing

A distinction is usually drawn between primary and secondary fish processing. Primary fish processing consists of gutting, filleting, freezing and other basic fish preparation. Secondary fish processing is the preparation of 'value added' food products from fish.

Primary fish processing is the first handling that takes place, usually in the fishing ports. Fish is often gutted at sea and filleted on shore, before moving to inland fish wholesalers and retailers. There are basically three sources of wet fish supply to the primary processing industry in the UK: landings by local vessels, overland fish from other UK ports, and various other sources of supply, such as foreign landings, container fish from Iceland and thawed frozen fish particularly from Iceland and the former Soviet Union. But often the main source of supply for primary fish processors is local landings. Two possible obstacles to getting into this business are (1) the modern hygiene standards that have to be met and (2) the fact that it is extremely competitive. Some of the crucial architectural skills are an ability to buy well, good control over staff and a network of potential buyers. Commercial acumen is important because people have to make large financial commitments quickly when they are taking part in a market as buyers.

Secondary fish processing

The secondary fish processing sub-sector is very adaptable and dynamic, but is dominated by large businesses. The demand for fish by secondary fish processors makes little distinction between fish purchased from local fish processors and imported blocks of fillets. Their main concern is that the specifications of the product in terms of species, quality and size should be exact and that the price should be right. Some large companies (such as Birds Eye Wall's) have now left primary fish processing altogether to concentrate on secondary fish processing – the preparation of products which are profitable in an expanding market. Birds Eye Wall's is part of a multinational, Unilever, and they see primary processing as a cost that they can escape. Their expertise is more profitably applied to efficient production of value added products and marketing. Nevertheless, this company still obtains some of its raw material from UK sources. Other companies who are involved in the preparation of value added products, such as Bluecrest, continue to value sourcing their supplies from regional wet fish markets.

The problem with fish processing is basically that at the top end of the business the standards are extremely high and require much investment to satisfy the buyers. However, niches for special products exist and there are still large numbers of successful small retailers who will be on the look-out for lower-cost options perhaps not produced to such exacting standards as for the multiples.

Fish retailing

Fish retailers face the same problem as other retailers – they are small and declining because of the competition from the multiples. Having said that, fish is one of those products where specialist knowledge is advantageous and this can take people a long way. So what does the potential business person need to look out for in a fishmonger?

As far as direct competition goes, multiples have been mentioned, but mobile fishmongers are also a source of problems to the shopkeeper, and because the investment requirement is low they keep springing up. A mobile retailer is an attractive point of entry into the business for the new entrant. Market stalls are also a source of competition to fishmongers with a shop, for much the same reason as mobiles.

When contemplating opening or buying a fish retail outlet, apart from the technicalities, points to consider include:

- The catchment area – population, unemployment, purchasing power, trends.
- Redevelopment, supermarkets, road widening, the availability of fish supplies; perhaps a friendly wholesaler
- The number of people passing the shop each day.
- The proportion of people passing who might want to buy some fish.
- The amount of money each of the potential buyers is likely to want to spend on fish or fish products.
- The reasons for the sale of the shop site by the previous owners.
- The gross margin – the difference between purchases of fish and sales of fish: figures which a new entrant needs to check against documentary evidence.

- Accounting information (sales, expenses, profits) is important, not just for 1 year.
- The value of the shop if the business fails.

Fish and chip shops

Much the same remarks can be made about fish and chip shops as for other kinds of retailer. These outlets are faced with severe competition from other fast food retailers, but it is striking how popular the dish still is – witness the queues of people lining up at key times of the day and week.

Fish restaurants

Again, this is a crowded, labour-intensive market in which a select few have been very successful.

The legal form of a business

The basic arguments for one legal form or another are much the same for new fish businesses as for any other small business. However, there is no doubt that making a good choice at this level is a fundamental strategic issue which must be examined in the light of the framework offered in earlier chapters. There are tax advantages and disadvantages for each form and there are no hard and fast rules. An accountant is needed to state the tax advantages of one form or another. Many people choose to set up a limited liability company, even if they are working on their own. An important issue for fishermen is the legal form adopted for their producers' organisations.

Sole trader

Provided a businessman keeps within the law he can simply start trading as a *sole trader*. Many people do so because there are no legal formalities, the person is totally responsible for his own life, he is taxed as an individual, he may start at any time and it is easy to wind up the business. However, there are disadvantages. Liability is unlimited. This means that in the event of liquidation the claims of people against the trader, such as money owed by the company to suppliers of goods and services, or debts owed to creditors, are unlimited. It can be a heart-breaking scene when a bankrupt sole trader loses everything. Sole traders usually trade under their own name. If they choose to trade under a name different from their own it should be registered with the Registrar of Companies (Companies House, Crown Way, Maindy, Cardiff, South Glamorgan CF4 3UZ, and in Scotland, the Registrar of Companies for Scotland, 102 George Street, Edinburgh EH2 3JD). The name and address of the trader should be displayed in all business premises, on letterheads, orders, invoices, receipts and payments.

Partnership

Partnerships are very common between accountants, solicitors, consultants or other groups of professionals selling their services. One reason for the attraction of this form of trading is that the group is not required to disclose trading information. The rules for operating partnerships are very much the same as for sole traders. All that is needed legally is for two or more people to agree to carry on a business together, intending to share the profits. A key defining characteristic of a partnership is that all the partners are personally liable for debts, even if those liabilities were incurred through mismanagement or dishonesty. A key advantage is that the partnership is assessed for tax purposes as a single entity so that if one part of the partnership is not doing so well, these losses can be offset against the more prosperous elements.

For these and other legal reasons most authorities recommend drawing up an agreement when starting a partnership. The agreement would cover matters such as mutual responsibilities, voting rights, policy decisions, withdrawing money from the partnership and accountancy procedures, membership, the period of the partnership (commencement, duration, termination), its name, activities and location, tenancy agreements and leases, capital investments (sharing and interest thereon), the profit sharing formula, drawings on account of profit shares, treatment of remuneration, e.g. fees and legacies, books of account, annual accounts including accounting date, clients' moneys, accountants, bankers, responsibilities of partners including holiday entitlement, negative covenants, motor cars, outgoing partners arrangements, compulsory retirement at a given age, admission of new partners, pensions, annuities, insurance, tax provisions and continuation election, covenants in restraint of trade, dissolution of partnership, expulsion of partners, serving of notices, partnership meetings and voting threat and a provision for arbitration in the event of disputes.

Limited liability company

A limited liability company is formed when two or more potential owners decide to set one up. The first step is the preparation of a Memorandum and Articles of Association. The limited company has a format to facilitate the aggregation of money to be used for production and trading. The people who supply the money, exchange it for shares in the company's capital and become *members* of the company. The members appoint directors to manage the company on their behalf and they retain the right to dismiss a director. The profits the company makes are available as dividends to the members. An important attraction of the limited company is that members are separated from management and also from the perils of mismanagement. Most companies of any size opt to become limited liability companies. Limited liability means that shareholders are not personally liable for any debts beyond the fully paid-up value of their shares. This applies even if the shareholders are working directors, unless the company has been trading unlawfully. In addition, the company's accounts must be audited by a certified or chartered accountant and records of the trading activities have to be filed annually at Companies House (address given under 'Sole Trader' above).

People form companies because they want their trading activities to be conducted under the auspices of an independent legal entity – to have its own legal personality in the jargon. Because companies have to abide by certain accounting and other practices most other businesses would prefer to work with a limited liability company rather than a sole trader. In other words, company status immediately advances a trader's reputation. Also, the company format is the simplest means of raising new capital through additional equity.

The name of a limited company is subject to approval of the Registrar of Companies. Directors must not use the same name as another company, a name which is criminally offensive or misleading, names used by qualifying entities (e.g. a bank), names connected with government or royalty, or implying national or international pre-eminence.

The normal practice is for a company to issue *shares* which have a *nominal* value. The nominal value has nothing to do with its real value on the stock exchange. It is not necessary to pay the company the whole of the nominal value of a share. When a company is wound up the shareholders are entitled to receive an amount proportionate to the shareholding after debts have been paid. This is known as 'equity'. They cannot be asked to pay more than the nominal value of their shares to meet the debts of the company; hence liability is *limited*. There are two categories of limited liability company, the public limited company (plc) and the private limited company (Ltd). The former requires an investment of more than £50 000, the shares can be traded and the reporting requirements are more onerous than for the latter.

Cooperative

A cooperative is owned and controlled by those working in it. Membership is usually open to all employees subject to qualifications laid down by members. One member has one vote irrespective of shareholdings. Profits are shared in accordance with agreed rules. In the UK, registration of cooperatives takes place under the Industrial and Provident Societies Act. A cooperative must have at least seven members and model rules are widely available from bodies such as the Plunkett Foundation.

In many forms of trading, cooperatives have lost ground in recent years. A recent highly topical example is the London International Financial Futures Exchange (LIFFE). Commentators now take the view that its mutual structure works to the disadvantage of its members because many of them are resistant to efficiency-enhancing change. In particular, computerised buying and selling of financial futures has so far been resisted although many of the big institutions who would like to move to lower cost computerised trading would welcome it. It seems likely that LIFFE will change to being a company so that it will have the flexibility to change as and when required.

Fish producers organisations

Most of the UK's fish producers organisations (FPOs) are cooperatives, but some (such as the Scottish Fishermen's Organisation Ltd and The Fish Producers Organisation Ltd) are companies limited by guarantee. Companies limited by guarantee have a group of subscribers who guarantee the debts of the company up to a certain amount. The advantage of this over the cooperative format is that the rules of operation can be made more flexible and crucially fix voting so that bigger companies in membership have a bigger say in outcomes (unlike the cooperative format with equality of voting).

Memorandum of Association

The Memorandum of Association includes:

- The company name
- Registered office location (England, Wales, Scotland or Northern Ireland)
- Company objectives – what it is in business to do
- The amount of nominal capital and how it is divided up into shares – thus showing the limits of shareholders' liability
- Number of authorised shares by type
- A declaration of association by at least two persons, called the *subscribers*.

Articles of Association

The Articles of Association include:

- Procedure for calling general and extraordinary general meetings
- Responsibilities and rights of directors
- Procedures for election of directors
- The company's borrowing policies
- Control of shares
- Voting rights of shareholders
- The role of the company secretary, his appointment and dismissal
- The appointment of auditors.

Responsibilities of directors

The application for registration includes a Declaration of Compliance which states that directors have complied with the Companies Acts. If the application is accepted by the Registrar of Companies a certificate of incorporation is issued. The certificate and registration date should be displayed on public view and the registration number and other formal details should appear on official company stationery. The Declaration of compliance is quite useful as it brings home to the directors of any company that they have certain legal obligations and, as a last resort, the courts may sentence them to prison if they fail to keep to them. They must be honest and diligent in attending to the interests of the company. They must not attempt to defraud shareholders or the public and they are in breach of their obligations if they allow the company to trade whilst it is insolvent. Indeed, if they do this they become liable for the debts of the company. They should have regard for the interests of employees and file accounts regularly. They should attend board meetings regularly.

The list of references at the end of this book includes a number of books by Adair (1985, 1987, 1988a, b) and also Nicholson's *How Do You Manage?* (1992). For directors, and those hoping to become directors, not familiar with these works and similar it is worthwhile examining what some of the gurus have to say. Nicholson is particularly helpful on leadership and team-building which are an important part of management.

The Role of the company secretary

The company secretary is a company's legal officer and has, therefore, statutory responsibilities to society as a whole as well as to his employers. Sometimes people who take on the company secretary's job do not have a clear conception of the majesty and importance of their responsibilities.

Responsibilities

- The company secretary serves as secretary of the corporation in accordance with its by-laws, Articles and Memorandum of Association.
- He arranges for meetings of the board of directors and keeps minutes of such meetings.
- He signs corporate documents and affixes the seal.
- He deals with company correspondence as the legal spokesman for the company.
- He keeps all the registers and other official documents as required by law. These include:
  - register of directors and secretaries
  - register of charges (legal obligations) and copies of the instruments creating the charges
  - minute books of general meetings
  - other books of minutes
  - register of members
  - register of debenture holders

- register of directors' interests
- register of substantial individual interests.

Company secretaries are the legal spokesman for their employer. They therefore have public as well as private responsibilities. They must, for example, ensure that all legal requirements for giving notice of meetings are met. They have to ensure that their companies follow the law on employment and dismissal of staff. For example, if a member of staff responsible for all insurances becomes redundant, his duties have to be assigned to someone else. It sounds obvious but is easily overlooked, as this author, when he was a company secretary, found out to his cost. A fisherman was injured at sea, but because the injury was not reported within the required time period the insurance company threatened to refuse payment. Fortunately the insurers did relent, but it was an alarming moment. In typical fish businesses, company secretaries also often have accounting responsibilities, so they must ensure that all money due to the company is collected.

Routine duties

- Accumulate items for decision at next board meeting.
- Circulate minutes of previous meeting.
- Book a room for the meeting.
- Draft and circulate an agenda in advance of the meeting – attach relevant papers (letters, reports, etc.) to the agenda.
- Prepare for the meeting – take paper, pencils, attendance book, minute book, a copy of the Memorandum and Articles of Association, spare copies of the agenda and papers, notes for the chairman.
- Get the minutes of the previous meeting approved by the board.
- Take notes.
- Write up minutes.

Issue and transfer of shares

The Memorandum of Association of a company states the number of shares and their nominal value that a company can sell (i.e. issue). The first issue is to the subscribers who must take at least one share; they become the company's first members and become so before their names are entered on the register. The transfer of shares is governed by law and must be registered. There are also various provisions for when shareholders die, etc.

Conduct of meetings

Shareholders' meetings

The Annual General Meeting (AGM) is a formal opportunity for shareholders to review the company's progress. The issues that must be considered are declaring a dividend, consideration of the report and accounts, the election of directors and appointment and remuneration of auditors. Extraordinary General Meetings are held for special business that cannot wait until the next AGM, for example, if there is an urgent need to amend the Articles of Association, or issue some new shares, or wind up the company.

Notice of General Meetings

The company secretary must send out the notice for General Meetings before the latest date stipulated in the Articles of Association. A typical agenda would include the following details:

- the name of the company
- the type of meeting
- date, time and place of meeting
- business to be conducted
- exact wording of any special resolution
- the authority by which notice is issued
- the name of the person issuing the notice
- note that members may appoint a proxy to vote for them
- date of notice.

Directors' meetings

The company secretary must send out an agenda. The quorum (minimum number of members in a board of directors required to be present before any valid business can be transacted) will be set out in the Articles of Association. Minutes need to be maintained.

First board meeting after incorporation

The first board meeting after incorporation is a special occasion. The prospective company secretary should ensure that the following items are dealt with:

- a report on incorporation. The certificate of incorporation and Memorandum and Articles of Association should be available for inspection by the members of the board
- a report on who the directors are
- appointment of chairman
- appointment of secretary
- appointment of auditors
- agreement on the registered office
- appointment of bankers and opening of bank account
- adoption of common seal

- disclosure of directors' interests
- any other business (AOB)
- date for next meeting.

Employing people

This text does not include an extended discussion of the management of human resources. However, it does acknowledge that this is an area of real challenge for any small business, and it is also where people can be tripped up if they do not pay careful attention to the legal rights of their employees. This caution applies with growing force in the countries of the EU.

In the UK if people work for more than 16 hours per week they have the following statutory rights:

- Written particulars of terms of employment
- Wage or salary pay statement
- Notice of termination of employment
- Time off for public duties
- Provisions in connection with pregnancy, maternity and rights to return to work
- Guaranteed payments
- Payments when suspended on medical grounds
- Defined redundancy terms
- Defined dismissal terms
- Statutory maternity pay (SMP)
- Statutory sick pay (SSP)
- Health and safety at work.

There are moves currently to extend employment rights to people who work less than 16 hours. Ludlow and Panton's *The Essence of Successful Staff Selection* (1991) is a useful text for those seeking advice on successful staff selection.

The fishing industry is not renowned for providing people with secure long-term employment. Most crews throughout the world are employed casually on a trip-by-trip basis. They are also generally paid a share of the value of the catch rather than a basic wage. The normal, although not universal, pattern is a 50:50 division of the value of the catch between the owner and crew. Of the crew's share, the skipper usually receives more than other crew members, and a junior trainee rather less.

Legal frameworks in other countries

Setting up a business in other countries is invariably different from the UK and businessmen setting up elsewhere have to be very careful to meet the legal requirements of others. There will always be competitors who want to see the new business venture go down if the letter of the law is not followed. Summaries of the state of play in Romania, Bulgaria and Poland follow (Poplawski, 1991; Frydman *et al.*, 1993). The examples are included for two reasons. One is that the author has worked in private sector fisheries development in each of the three countries so has had some exposure to the problems. The second is that all the countries listed follow a continental rather than an Anglo-Saxon model.

This latter point is very significant, especially with respect to accountancy matters. European countries still follow different accounting practices. There is a growing preference for Anglo-Saxon approaches which are widely accepted as being more transparent, particularly from the point of view of shareholders, than the continental model. The EU is trying to establish common practices and this will no doubt be achieved in time.

Romania, Bulgaria and Poland

There are subtle differences between the legal frameworks for companies in each of the three countries Romania, Bulgaria and Poland. However, they are broadly similar, with each country systematically opening itself up to foreign ownership and freedom to trade (Poplawski, 1991; Frydman *et al.*, 1993). Each country has six categories of trader:

(1)   State-owned companies. They are akin to the old nationalised industry boards in the UK, such as the Central Electricity Generating Board or the National Coal Board. They govern the main strategic industries, such as defence manufacture, and they are financially autonomous.
(2)   Cooperatives are part of the old communist legal framework and operate in agriculture. They are now almost defunct and discredited.
(3)   Private merchants or traders.
(4)   Partnerships.
(5)   Limited liability companies – designed for smaller companies with more limited reporting requirements. State-owned (single owner) commercial companies are established for many manufacturing processes, including fish processing and aquaculture (another hangover from the communist era), but eastern European countries are making major efforts to sell off these state holdings to the private sector.
(6)   Joint-stock companies – most like plcs in the UK. They have a minimum capitalisation, and, except for state-owned companies, a minimum number of shareholders (e.g. five in Romania). Liability is limited.

The Bulgarian Commercial Code was adopted in 1991. It divides all economic agents into 'merchants' and 'non-merchants'. A merchant is someone involved professionally in business activities, but the category excludes farmers, artisans, members of the professions and self-employed providers of services. Cooperatives are governed by a separate law (the Law on Cooperatives, 1991). Under the Law on Activity of Foreign Persons and Protection of Foreign Investments (1992) foreigners are governed by the same legal principles as apply to Bulgarians, except where specific regulations provide otherwise. In general, foreign businesses do not need special permission to invest. They have the right to repatriate their investments and profits if they wish. Ownership of land is excluded from the rights of foreigners. If a foreigner wants to invest in Bulgaria he does so under the Commercial Code.

Polish company law is similar to Bulgaria. The current round of simplifications are listed in *The Polish Commercial Code*, as published by the Polish Bar Foundation (Poplawski, 1991). The basic structure of smaller limited liability companies and larger joint stock companies is similar to Romania and Bulgaria.

Interpretation of accounts

The following example is included to assist in the interpretation of business plans. It is adapted from Brett (1995).

Assume Fishy Business is a young company selling processed fish in units of 1 kg and was set up one year ago by eight associates. Each put £5000 into the business by subscribing 5000 £1 ordinary shares and they borrowed the rest of the money required to set up the business. They put in £40 000 in equity and borrowed £30 000 as a long-term loan and £8000 as an overdraft. Table 6.1 shows the balance sheet and profit and loss account of Fishy Business.

Fixed and Current Assets

Assets are what the company *owns*. They are known as *fixed* if they are not something the company is buying and selling or processing in the course of trade. The main fixed assets of Fishy Business are buildings and plant. In the case of this example the company owns £27 000 worth of fish processing equipment. Originally Fishy Business paid £30 000 for the equipment, but has reduced its value by £3000 because of *depreciation*. £27 000 is the *book value* of the equipment. Current assets, otherwise known as *working capital*, are assets on the move. These include stocks of raw materials (fish, packaging and other inputs), finished product, money owing to the business (debtors) and money held in the bank.

Liabilities

Liabilities are what the company *owes*. Current liabilities are short-term debt to the creditors of the business if the amounts are due within a year. They are deducted from current assets to give net current assets, sometimes known as net working capital. They include trade creditors (the value of goods purchased but not yet paid for), tax payable, the proposed dividend and the overdraft. The ratio of

Table 6.1 Example of a balance sheet and profit and loss account.

Fishy Business Ltd	
Balance sheet for the year to 31 December	
	£
Fixed assets	
Plant and machinery	27 000
Current assets	
Stocks	50 000
Debtors	35 000
Cash at bank	6 000
Total current assets	**91 000**
Current liabilities	
(Creditors: amounts due within 1 year)	
Trade creditors	20 000
Tax payable	4 200
Dividend proposed	4 000
Bank overdrafts	8 000
Total current liabilities	**36 200**
Net current assets (Current assets minus Current liabilities)	**54 800**
Total assets less Current liabilities	**81 800**
(Creditors: amounts due in more than 1 year)	
Term loans	30 000
Net assets (£81 800–£30 000)	**51 800**
Represented by:	
Share capital (£1 ordinary shares)	40 000
Revenue reserves	11 800
Shareholders' funds	**51 800**
Profit and loss account for the year to 31 December	
	£
Turnover	200 000
Trading profit	24 000
Less	
Interest paid	4 000
Profit before tax (pre-tax profit)	20 000
Less	
Corporation tax	4 200
Profit after tax	15 800
Less	
Dividends	4 000
Retained profit (into the balance sheet as Revenue reserves)	11 800

current assets to current liabilities is often regarded as important because, if it is greater than 1, it tells investors that the business can meet its current debts in an emergency. Longer-term debt might be a long-term loan that the shareholders have taken out to set up their business. In this case we imagine that there is a £30 000 term loan taken out to pay for the equipment.

Net assets

Net assets are the residual value of the business after the deduction of all debts. The total assets are worth £118 000. This is the value of stocks (£50 000) plus debtors (money owed to the business – £35 000) plus cash at the bank (£6000) plus the depreciated value of equipment (£27 000). Deduct from this borrowings of £38 000 (term loan of £30 000 and overdraft of £8000) and also deduct other liabilities of £28 200 (£20 000 plus £4200 plus £4000) and it leaves £51 800 as net assets, or what the business is worth to its shareholders. Balance sheets must balance so these net assets of £51 800 are attributed to the shareholders in the balance sheet. This value consists of the £40 000 they put in at the beginning plus another £11 800 in reserves.

Tax

Suppose that £4000 is paid as dividends divided equally among 40 000 £1 shares issued. The dividend is often expressed as an amount in pence per share; thus in this case it is 10p *net* per ordinary share. This is equivalent 12.66p before tax since income tax of 21% is already assumed to have been deducted; 12.66p is the *gross* dividend per share. Most yields are usually expressed in the financial press as *gross*, i.e. before tax.

Investment ratios

Balance sheet gearing

The ratio of borrowed money (debt) to shareholders' money is known as the gearing ratio. A highly geared company has a high ratio of borrowed money to equity (in American terminology 'the leverage is high'). If a company is doing well, high gearing is good for shareholders because the burden of interest out of total profits is low, but if profits are low or negative the interest burden resulting from high gearing worsens the situation.

The total borrowings of Fishy Business are £8000 bank overdraft and £30 000 term loan, i.e. £38 000 in total. Trade creditors are not included because they have no impact upon interest charges. Shareholders' funds are £51 800. The most common way of relating them is to calculate borrowings as a percentage of shareholders' funds to give the gearing ratio, in this case 73.4%. Sometimes the balance sheet gearing ratio is calculated on net borrowings (borrowings less cash – £32 000 after £6000 cash in the bank has been deducted). The ratio is then 61.8%.

Income gearing

If £4000 is the interest payment and the trading (or operating) profit is £24 000, the income gearing is 16.7% (4000/24000 × 100). This is a low gearing ratio because the interest payments are low for a profitable company. But suppose a different company has a trading profit of £100 000 with interest payments of £50 000 leaving a pre-tax profit of £50 000. The gearing ratio is therefore 50%, much higher than that of Fishy Business. If the trading profit then increases by 50% to £150 000 the interest payment of £50 000 leaves a pre-tax profit of £100 000. So for an increase of 50% in trading profit, pre-tax profits rise by 100%. However, if there is a 50% fall in trading profit, nothing is left after £50 000 interest has been deducted.

The appropriate gearing ratio varies between companies and sectors. High gearing might be appropriate for companies with a stable and rising income, such as a business deriving rents on good commercial buildings. However, for companies in a more uncertain environment where profits can go up or down a high gearing ratio makes them vulnerable.

Earnings Per Share (Eps)

Earnings per share (eps) are calculated by dividing the net profit (profit after tax) by the number of shares. The 40 000 shares of Fishy Business earn £15 800 (dividends plus retained profit) or 1.58 million pence, equal to 39.5p per share. Suppose Fishy Business decides to take over another company exactly similar to itself in exchange for shares. It creates 40 000 new shares and gives them to the owners of the company to be taken over. The newly enlarged Fishy Business now has combined net profits of £31 600, exactly double the size of the profits of the old business. Shareholders, however, have not gained because the eps is the same. The lesson here is that growth through acquisition does not, in itself, add value to shareholder returns.

Dividends per share

Fishy Business pays out dividends of £4000. Expressed as an amount per share this is 10p (£4000 = 400 000p, divided among 40 000 shares). This is the net dividend per share because income tax at the basic rate of 21p in the pound is deemed to have been paid. To find the gross dividend the net dividend needs to be grossed up. The calculation is 10p divided by 0.79 = 12.66p. The gross dividend is the one on which yield calculations are normally based.

Dividend cover

The directors of Fishy Business recommend to shareholders the proportion of profit to be paid out as dividend. In this case, £4000 is paid out on a net profit of £15 800; therefore the dividend cover is 3.95 (£15 800 divided by £4000).

Stockmarket ratings

Assume that the current market price of Fishy Business shares is 300p. The *yield* for an investor who purchased shares at 300p is 4.22% of his payment (12.66p divided by 300p × 100). This is the current yield on shares in the business and as this is a fairly low yield investors might accept this if it is a fast-growing company. The problem is that the dividend is arbitrary – it could be set higher or lower by the same company with the same profits. For example, the directors might decide to pay out twice as much profit in dividends; then the dividend would be 27.4p gross and the yield is then 8.44%. In other words, the yield does not tell us much about the underlying strengths or weaknesses of the company.

Price-earnings (PE) ratio

The price-earnings (PE) ratio is a relative measure of how highly investors value the earnings a company produces. It is derived by dividing the eps into the market price of shares. If the eps of Fishy Business is 39.5p and the market price of shares is 300p then the PE ratio is 7.6 (300 divided by 32.5). A share with a high PE ratio is one that is currently valued highly by investors. They expect income from the shares to rise in the future so companies with high PE ratios are the fast-growing companies. A company operating in a more static or higher risk market would probably have a lower PE ratio. When figures are quoted 'xd' this means ex-dividend and the buyer does not acquire right to a recently announced dividend. When figures are quoted 'cum div' or 'cum dividend' this means that figures are quoted on the assumption that the buyer acquires the right to the dividend.

Net assets per share

The shareholders' funds in Fishy Business amount to £51 800; this is the book value of shareholders' interest in the company. This can be expressed as net asset value (NAV) per share: £51 800 divided by 40 000 = £1.295 (129.5p). The NAV is always based on book values. Shares are commonly described as standing at a premium to (i.e. more than) or at a discount to (i.e. less than) the NAV.

Return on assets

The return on assets is normally expressed as the profit before tax and before interest on long-term loans expressed as a percentage of shareholders' funds plus long-term loans, deferred tax and minority interests.

Summary

Options for fish businesses

The fish industry has many different subsectors. Some of the options for investment in the industry include:

- The catching sector
- Fish farming
- Primary fish processing
- Secondary fish processing
- Fish retailing
- Fish and chip shops
- Fish restaurants.

The legal form of a business

Sole trader

Provided a businessman keeps within the law he can simply start trading as a sole trader. Many people become sole traders because there are no legal formalities, the person is totally responsible for his own life, he is taxed as an individual, he may start at any time and it is easy to wind up the business.

Partnership

Partnerships are very common between accountants, solicitors, consultants or other groups of professionals selling their services. A key defining characteristic of a partnership is that all the partners are personally liable for debts, even if those liabilities were incurred through mismanagement or dishonesty. One of the main attractions of partnerships for professionals is that they can escape disclosure of commercial information. A partnership agreement is advisable.

Limited liability company

A limited liability company is formed when two or more potential owners decide to set one up. The first step is the preparation of a Memorandum and Articles of Association. The limited liability company has a format to facilitate the aggregation of money to be used for production and trading. The people who supply the money exchange it for shares in the company's capital and become *members* of the company. The members appoint directors to manage the company on their behalf and they retain the right to dismiss a director. The profits the company makes are available as dividends to the members. An important attraction of the limited company is that members are separated from management and also from the perils of mismanagement.

Cooperative

A cooperative is owned and controlled by those working in it. Membership is usually open to all employees subject to qualifications laid down by members.

One member has one vote irrespective of shareholdings. Profits are shared in accordance with agreed rules. In the UK, registration of cooperatives takes place under the Industrial and Provident Societies Act. They must have at least seven members and model rules are widely available from bodies such as the Plunkett Foundation.

Fish producers' organisations

Most of the UK's fish producers' organisations (FPOs) are cooperatives, but some (such as the Scottish Fishermen's Organisation and the Fish Producers' Organisation) are companies limited by guarantee. Companies limited by guarantee have a group of subscribers who guarantee the debts of the company up to a certain amount. The advantage of this over the cooperative format is that the rules of operation can be made more flexible and crucially fix voting so that larger companies in membership have a larger say in outcomes (unlike the cooperative format with equality of voting).

Responsibilities of directors

It is valuable to bring home to the directors of any company that they have certain legal obligations and, as a last resort, the courts may sentence them to prison if they fail to keep to them. They must be honest and diligent in attending to the interests of the company. They must not attempt to defraud shareholders or the public and they are in breach of their obligations if they allow the company to trade whilst it is insolvent. Indeed, if they do this they become liable for the debts of the company. They should have regard for the interests of employees and file accounts regularly. They should also attend board meetings regularly.

Role of the company secretary

The company secretary is a company's legal officer and has, therefore, responsibilities to society as a whole as well as to his employers.

Legal frameworks in other countries

Setting up a business in other countries is invariably different from the UK and businessmen have to be very careful to meet the legal requirements of others. There will always be competitors who want to see the new business venture go down if the letter of the law is not followed. Romania, Bulgaria and Poland follow similar continental models.

Interpretation of accounts

The balance sheet is the key business account. It states what the company owns, its assets, and what the company owes, its liabilities. Each year the money gained or lost to the business recorded in the profit and loss account feeds back into the balance sheet to expand or contract the business.

Chapter 7
The Business Plan in Fisheries Development

The Basics

Planning

The business plan is a key document. It forces the established business or the new entrant to think through what he is going to spend on costs and how those costs are going to be recovered and contribute to profit. The plan should identify distinctive capabilities and show how these can reasonably be expected to translate into competitive advantages and economic added value. Unless the business can do this, applying for finance is a waste of time and money.

Pride of place in the business plan goes to cash flow because, without cash, there is no business. The profit and loss account is different from cash flow because it also includes changes in the values of assets. Indeed, a business can be profitable but still go to the wall because the profits are due to changes in asset values (such as property prices) which are not converted into cash flow. Similarly, a company can make losses but have a positive net cash flow if the value of its assets has declined but its cash income exceeds its cash expenditure. However, the fundamental business accounting document is not the cash flow statement, nor the profit and loss account, but the *balance sheet* which states quite simply what a business owns and what it owes. If it owes more that it owns it is in trouble.

The role of information

It is perhaps worth stating that there is much information available to business-people in the UK, although probably not many are aware of what can be accessed. Generally, in developed countries a search can reveal such information, but the developing world is a different matter. In countries with weak public information systems it takes a certain amount of intuition to get to know what is going on.

A selection of UK sources of information about the fish industry follows:

- *Annual Census of Production – PA 15.2 Processing and Preserving of Fish and Fish Products* (Office for National Statistics, annual). Includes value added estimates.

- *The Complete Small Business Guide: Sources of Information of New and Small Business* (Barrow, 1995).
- *A Workforce Audit of the Sea Fish Industry in the UK* (Cemare, 1991). CEMARE, the Centre for Marine Resource Economics at the University of Portsmouth, has its own library. This contains a comprehensive collection of published and unpublished fisheries-related material of which this report is only an example.
- *Return on Capital in the European Fishery Industry* (Davidse et al., 1997).
- Digest of Welsh Statistics. Includes fisheries data. Published annually by the Welsh Office Publications Unit, Cardiff.
- *A Report for the Hull Economic Development Agency Food Sector Initiative* (Devos, 1995).
- The Environment Agency. Has details of salmonid fisheries and river fisheries.
- ESRC (Economic and Social Research Council) Data Archive, at the University of Essex. Includes UK Census data, General Household Surveys, Family Expenditure Surveys and the Labour Force Survey.
- Family Resources Survey. Conducted by the Department of Social Security. Estimates are available for the standard regions. Available from the ESRC Data Archive at the University of Essex.
- Training and Enterprise Councils (TECs). For example, the Humberside Training and Enterprise Council publishes the very well-informed *Humberside Economic Review*, a model replicated by many TECs which are distributed throughout the country.
- Kompass, the Authority on British Industry 1997/1998. Supplies information on companies, products and services supplied and financial data on each company.
- Labour Force Survey. The Office for National Statistics (ONS) is responsible for this. Data are collected from a sample of around 63 000 households. The data concern the economic activity of those aged 16 and over their employment (employees, self-employed, government-employed and on training programmes, unpaid family workers), occupations, industry worked in, job-related training, hours of work and unemployment (International Labour Organisation definition), length of time without a job, redundancies, those economically inactive and the reasons, income and demographic characteristics. It covers all the standard regions and counties. Estimates are available for local authorities and TEC areas. Since 1992 it has been a quarterly survey and available through NOMIS (National On-line Manpower Information Service).
- National Food Survey. A Ministry of Agriculture, Fisheries and Food (MAFF) study of 8000 households, conducted quarterly.
- NOMIS. A database of labour statistics run on behalf of the ONS by the University of Durham. Includes a comprehensive range of official statistics relating to the labour market, including employment, unemployment, earnings, vacancies, training, the Labour Force Survey and the Census of Population. Data can be extracted down to quite small units – ward and district level. Most of the TECs are on-line to NOMIS.

- New Earnings Survey. Supplies data on average earnings in the UK including gross weekly and hourly earnings and hours worked. Analyses are done by industry, region and age group.
- *The Fishing Industry of Yorkshire and Humberside: a Regional Study* (Palfreman *et al.*, 1993).
- *Guide to Official Statistics* (Office for National Statistics, 1996). A comprehensive reference tool acting as a signpost for all who need ready access to statistics. There is a wealth of economic and social information available from the Government Statistical Service and other official organisations throughout the UK. The ONS used to be known as the Central Statistical Office (CSO).
- Regional Accounts, including Economic Trends, Regional Trends and the GeoStat Dataset. Regional Trends includes detailed data down to NUTS 2 (Nomenclature of Territorial Units for Statistics, sub-regions). Available from the ONS.
- Register of Fish Farming and Shellfish Farming Businesses. Fish farms must be registered with MAFF who are prepared to supply aggregate data on these farms.
- Sea Fish Industry Authority. Various reports as well as a good library.
- Sea Fisheries Statistics and Scottish Sea Fisheries Statistics. In addition to monthly and annual publications, MAFF and the Scottish Office are able to supply landings by port, administrative district, species, area of capture, fishing organisation and port of registration. They are also able to supply details of fishing vessel registrations and numbers of full-time and part-time fishermen.
- The Rural Development Commission, Local Enterprise Agency, Small Firms Service, Small Business Clubs.
- Universities. For example, the University of Hull Information Services has CD-ROMs, a European Documentation Centre, an extensive reference and statistics section and World Wide Web access.
- Public libraries. Always a wealth of information, especially for local statistics.

But Tom Peters is more sceptical

Tom Peters, the well-known management guru, is more sceptical. The following typical Tom Peters, quotation might lead a prospective businessman to think that it is impossible to develop a picture of what is happening, but it also might be an idea to give the question as much thought as possible.

'But what about facts – cold, hard statistics? Guess what? There aren't any. Been following the US health care debate? The principal players can't even agree on how many of us are uninsured – estimates vary by millions. Ditto the new jobs debate: some confidently proclaim, with (literally) a ton of supporting evidence, that most new jobs pay well; others confidently point to wage slaves for most of the new positions.... Business is poetry.... I know that anything I come across has at least 100 plausible explanations; moreover anyone can produce convincing evidence that will completely negate the hard data I'm now devouring.'

The above is quoted from an article by Tom Peters in *The Independent on Sunday* (Peters, 1994). See also *Liberation Management* (Peters, 1992) for a readable and refreshing alternative view.

Stages of business inception

Whether it is an entirely new business or a new project within an existing business, getting the business started follows a similar pattern. Chapter 14 includes an outline of the World Bank project cycle. The following is the normal process of new business development, which, on inspection, is found to be very similar to the World Bank project cycle.

(1)   Identifying a business concept and then doing a preliminary study is the first stage. Sometimes, following Tom Peters, ideas come from the intuition of the entrepreneur. However, the fishing industry is well established and opportunities are far more likely to emerge from a more careful process of research to assess where market segments are expanding. It is worth recalling Chapters 2 to 5; the fish industry business has to look to its own strengths – its distinctive capabilities – to find new opportunities.

(2)   The second stage is the more detailed preparation of the project. If possible it is a good idea to try to prove the business idea, especially if one is seeking external finance. This is often not possible so a more detailed and well-researched assessment of feasibility is required – this is the heart of the business plan.

(3)   The third stage is implementation or lift-off; this is when time and money are really absorbed, because the business has to start spending on the hardware requirements. Premises have to be obtained, staff employed, business names and brand names registered, professional advisers may need to be engaged and publicity must be prepared.

(4)   Finally, the business or business project needs to be managed or monitored to ensure that it stays on course. One must keep track of results, comparing performance with the business plan. Changes in the market and the law, especially hygiene standards, need to be followed. The architecture of workforce, customers and suppliers needs to be maintained. Problems need to be anticipated and products improved. The look of the premises needs to be monitored and brought up to date. Observing the successes and failures of competitors is crucial.

Reasons for the business plan

A business plan is essential in applications for finance. If the owner intends to apply to a funding agency, or a bank, venture capitalist, or to family and friends he needs to think through the basic question of where he expects income to come from and how he expects to spend the money coming in. A business plan is also a logical framework for people who work in the business. It shows owners and employees where the business is expected to go. Of course, with a small business it is widely expected also that if a new opportunity arises then the business might pursue it even if it is not in the plan.

A business plan also draws attention to weaknesses in the company, areas where inadequate attention has been given. A good example would be the lack of attention to issues of food hygiene in the context of more and more public and private pressure on food businesses. A business plan is especially useful in exposing weaknesses in the existing management system because it forces the directors to think hard about the way things are done and what improvements might be made. A business plan identifies areas where the company needs to develop new expertise. It is a way of assessing whether a new product, process or investment will add to or detract from the value of the business.

The demands of business planning

To most people operating a small business, the thought of spending time on preparing a business plan is unwelcome. When confronted with the choice between making a sale, sorting out a production problem or writing a business plan most people in the fish sector (and other industrial sectors too) avoid the paperwork. However, provided it is kept simple, it is not impossible to do. Moreover, there is plenty of software available to help the businessman with the number-crunching aspect of it.

What is a business plan?

A business plan is an orderly statement by the directors of a company of how they expect their business to change in the next few years. Notice that this definition refers to *change*, not necessarily *growth*. There are circumstances when it is logical for a business to engage in a well-managed contraction (as described in Chapter 3), so that it divests itself of its less profitable or loss-making components. Or it may be that the business is sound, but it still makes sense to break it up so that part or all of it can be sold. This can often happen when the owner wishes to retire, and he can receive a better deal for parts of the business than for the whole. Notice also that the statement does not specify a time period. A business plan covering 2 or 3 years will be enough for many small businesses, especially in the service sector, but a business plan for a company which intends to make larger investments should cover a longer period, say up to 10 years ahead.

Main issues in a business plan

The business plan should include information under each of the following headings. As will become clear in the examples that follow, for practical reasons it is sometimes not convenient to follow the exact pattern set out below. This is a guide to the issues that need to be covered rather than a model to be followed to the letter. A model can be found in Barrow *et al.* (1992; pp.240–242). Some summaries of fish industry business plans are included in the Appendix to this chapter.

Executive summary

(1) History of the business

The plan should state what the business is all about – the business mission. The business plan should supply a summary of the trading record and activities of the business since inception, or during the last 5 years, whichever is the shorter. The number and categories of employees throughout the period should be noted. The registered office, bankers, auditors/accountant, solicitor and company registration numbers should all be included in the introductory section. The introduction might also include a statement of the business mission.

(2) Shareholders and senior managers

The plan should state the ownership structure and the senior management of the business. The former is important because it supplies potential financiers with an understanding of the commitment they might expect from the business. For example, if a fish farm in eastern Europe is owned by the state, it will not expect the same commitment to profit compared to the situation if the owner is a private individual. Nowadays, financiers might also feel considerably less secure about total or even partial state ownership of assets. The quality of senior management is a crucial piece of the jigsaw too. Potential financiers want to know the track record of the senior management. Firstly, they will wish to be able to make a judgement on the ethical standards of the applicant for a loan, and also they will want to know if the management has the capability to sustain the business. It may be appropriate to include the CVs of senior personnel as an appendix to the plan.

(3) The marketplace

Fish marketing is discussed in Chapter 8. The business plan needs to review and interpret the data that have been collected. The main questions concern the market according to geography, customers and sector. The plan should aim to identify where possible the total market potential, the current existing market and target share. Competitors should be identified, paying attention to pricing, product improvement and performance. The first step is to identify the market segments.

(a) *Market segments:* Fish business might be selling to industrial or consumer markets. Fish farmers or fishing vessels usually sell direct to wholesalers – the food industry's equivalent to industrial buyers. Restaurateurs and fish retailers sell to consumer markets.

For consumer markets the income and other family characteristics of potential buyers will be significant. For example, it may be possible to identify more than one market segment for a fish and chip shop: (1) day-time buyers for the family and (2) buyers as people leave the pub. This may then point the way to offering subtly different products to the two segments.

Geography is important. Fish-buying tastes differ between locations. The haddock/cod differentiation for different parts of the UK is well-known. Another example is the taste for unsalted dried kapenta (a freshwater sardine from Lake Kariba) on the northern, Zambian shore whilst the Zimbabweans prefer their dried fish to be salted.

Social and psychological factors need to be thought through. It may be possible to develop some interesting ways of differentiating between potential buyers of fish on the basis of their social class. There is not much point in locating an expensive up-market fish restaurant in down-town Hull, but there may be a market for a carefully planned fish and chip shop differentiating itself by offering skinless fish as compared to the traditional skin-on variety.

What about frequent or infrequent buyers of fish as a basis for offering different products? Fish restaurants may be able to develop new and interesting offerings to the gourmets, but simpler, more well-known fish may be more suitable for less frequent buyers.

Price will be important. The example of skinless fried fish has been noted, but this will be more expensive than skin-on because of the implied loss of weight from the removal of the skin. For the same amount of fish by weight the consumer will have to pay more for skin-off. This may or may not put the product outside the segment's price range.

Distribution methods are also a possible source of segmentation. The potential entrepreneur may be able to think of a novel way of distributing the product (such as home deliveries) to expand sales.

The same kind of thought processes are required for industrial market segmentation. Geography, customer requirements, market shares, frequency of deliveries, may all help the businessman to consider how his market might be divided up.

(b) *Product attributes:* The businessman needs to ask why the customers will buy his fish. Issues include reputation, appearance, delivery times and waiting times, and quality. The product then has to be adjusted to meet the objective requirements of the market.

(c) *Forecasting sales:* The final marketing question to be addressed in the business plan is to try to quantify what the sales are likely to be, bearing in mind the preceding market analysis. An estimate of the total market size is the starting point. The expected share of the market for the business can then be estimated.

(4) Product and/or services

The business plan should identify the current range of products and services offered, including the unique selling features and selling benefits associated with each one. This section should be developed by a discussion of the potential for adding value to existing products and services and for developing totally new products and services.

(5) Business strategy

Business strategy is discussed in Chapters 2 to 5. The business plan should describe the strategic thinking of the senior management. There are a number of issues to consider: the capital structure and financing, the development of products and markets, hygiene regulations, the production technology and equipment, the distribution system, relationships with suppliers, premises, future management structure and international aspects (trade and finance).

(6) Shorter-term objectives

The immediate issue for the business is what it is going to do over the next 12 months. The major assumptions about the conditions affecting the business should be stated.

(7) Management information system

Management information is likely to be important because it shows how well the business is controlled. Areas that are important include credit and credit control, research and development, purchases, production costs, personnel costs, cash flow, sales and petty cash.

(8) Financial requirements

This section should include a discussion of and tables showing the financial needs of the business. It would include the timing of intended expenditures, sizes, purpose and planned benefits to business from these. The section would also show the contribution to future investment requirements to be made by shareholders and other sources of capital. It would also indicate the plans for the repayment of the investment.

(9) Financial information

Audited accounts for up to 5 years and interim figures where audited accounts are not available should be included with the plan. An important component of the business plan is a cash flow forecast. Sometimes this is a quarterly or even monthly prediction for at least 12 months. Thereafter annual figures are usually sufficient. The cash flow forecast provides the basis for predicted profit and loss accounts and balance sheets for the next 2 or 3 years. Detailed advice on doing this is not included in this book. Barrow *et al*. (1992) supplies an excellent step-by-step guide.

(10) Security

Potential financiers want to be sure that their investments are secure. So any relevant information, such as charges over assets, assets of shareholders, partnership agreements, the status of directors' loans to the business, should be included. The section should also include the Memorandum and Articles of Association.

Basic rules of bank lending

Will the bank or other source of finance lend on the strength of a good business plan? The following is a list of the basic principles which commercial banks apply when assessing applicants:

- An equal contribution to the capital from the businessman and the bank is normally expected. This implies that applicants with very limited resources are often in difficulties in supplying the equity that the financier requires.
- Full security on bank's lending: The bank requires security to be easy to value, readily marketable and easy to obtain title to. It is often provided by personal guarantees, land and buildings, stocks and shares, life assurance policies, business assets such as debtors, stock and fixed assets.
- Capacity to service interest charges: The ratio of interest charges to profits (the income gearing ratio) expected by financiers might be about 1:4, although the requirement would vary from bank to bank (see Chapter 6).
- Capacity to repay loan: the financier would seek clear evidence that the project will generate the financial resources to repay the requested loan.
- The borrower himself: the bank looks for integrity and reliability as well as the evident capacity for carrying out the business plan.

Principal sources of investment finance

Commercial reality is that finding finance for good-quality projects is not really a problem. There are many sources of potential investment capital for projects in the UK and also in other countries. This is the result of a world in which there are many savers seeking high returns and many businesses looking for outlets for that money. The problem really is making sure the business works – which is much harder to achieve than raising the money.

Development agencies

The fish industry has traditionally been relatively well-supported by development agencies. So the first line of enquiry that any fish industry entrepreneur should make is to government offices and development banks to investigate what might be available.

Self, family and friends

Finance from the proprietor himself, from his family and from friends might be in the form of loans or equity. The latter is when money is invested in the business in exchange for a share of the ownership.

Shareholders and directors

There may also be others who do not fall into the 'self, family and friends' category who are prepared to lend money to the business or invest in it.

Internal company sources (not applicable to start-ups)

For established businesses, retained profit is of fundamental importance as a source of investment finance, and also as a signal to potential new investors. Even if the business is not particularly profitable, cash in the bank is impressive. Sometimes payments to creditors can be delayed or cash can be raised on the security of stocks held. Property and insurance policies are particularly useful as security against which immediate cash requirements can be met.

Clearing banks

A 'clearing bank' is UK terminology for the normal commercial banks, most familiar to the general public for their retail services. Nowadays, they offer a whole range of financial services, such as foreign exchange dealing and corporate finance. However, they also specialise in providing loan finance for new businesses. Some building societies also offer business development loans.

Specialist financial institutions

'Investment banks' are the American equivalent of British 'merchant banks' and offer specialist financial services, more to larger than smaller businesses. There are a number of other specialist financial institutions, such as pension funds and insurance companies, operating in the UK which also offer financial services to industry.

Grant/loan awarding institutions

The Sea Fish Industry Authority has a long and honorable history of supplying grants and loans for the construction of fishing vessels. It no longer does this, but in some parts of the country grant and loan assistance can be obtained through various schemes. In the UK the first point of call for further information are regional offices of the Department of Trade and Industry, MAFF and the Scottish Office. The Pesca initiative for the fish industry is a recent development. Pesca is described in Chapter 11. Williams (1998) supplies comprehensive information about general UK investment finance. Eastfish, the FAO organisation based in Copenhagen, has details of the investment finance available for fisheries development in Europe.

Venture capital

Venture capital is increasingly available in the UK. 'Investors in Industry' (3i) is the leading player in this market with stakes in over 5000 businesses. There are also numerous private individuals ('business angels') who will be prepared to support the development of a business.

Mergers and acquisitions

An important source of new finance is when another company buys into, or buys outright, the business in need of finance. Very frequently the only realistic means of expansion is to get another party involved.

Second-tier finance

Hire purchase

Under hire purchase schemes the buyer of a piece of capital equipment escapes heavy initial costs by making regular payments resulting in the ownership of equipment at the end of the hire period. Under UK tax law, capital allowances and other grants may be claimed at the outset by the hirer. A deposit is required and the calculated interest may be tax deductible (again, under UK law). When no initial deposit is payable by the hirer the agreement may be called lease purchase.

Leasing

Leasing capital equipment is another means of obtaining it. The capital allowances and grants are claimed by the leasing company and then, either wholly or partly, passed on as reduced rentals. At the end of the lease period a secondary period of lease may be taken up as an option at a nominal rent. Alternatively, there may be an option to purchase the equipment at the end of the lease period.

Lease rentals are treated as expenses and may, therefore, be deductible against tax. 'Closed ended leases' run for a fixed term of 1 to 5 years. 'Open ended leases' can be terminated at any time by either party after the nominated minimum period has passed. 'Balloon leases' allow some of the payments to be made in a lump sum at the end of the lease agreement.

Leasing, as opposed to buying, capital equipment has advantages: it reduces the capital needed for start-up or expansion, it is off the balance sheet, therefore the ratios are unaffected, it is quick and easy to arrange, and there are tax advantages to the sole trader or partnership in the early trading periods. However, it is expensive, and the tax advantages may be limited and will certainly need to be investigated before entering an agreement.

Contract hire (otherwise known as operating leases)

With contract hiring, the hirer is responsible for maintenance and, if necessary, replacing the asset.

Sale and leaseback

Sale and leaseback is a way of obtaining cash. For example, property might be sold to a buyer who then agrees that the seller may continue to occupy it for an agreed period. Thus the buyer gets title to the property but releases cash.

Factoring and invoice discounting

A company needing cash can sell its business debts to a factoring company. The factoring service includes sales ledger, invoicing, insurance, cash collection and credit control systems. The reduced service, without the management element, is known as invoice discounting.

The principle of matching finance

The principle of matching finance is that money raised by a business should match the purpose to which it is applied. For example, a common error is to use an overdraft facility to purchase plant and equipment. Then if the overdraft is called in by the bank the business faces a cash flow problem. If finance matches its purpose the risk is contained.

Equity is core finance and permanent capital, to set the business up and keep it going. Typical sources are self, family shareholders, retained profits, trusts, private investors, development corporations and boards, venture capital funds, merchant banks, insurance companies, pension funds and business competitions.

Short-term funding (0 to 3 years) is for short-term working capital, seasonal requirements and bridging finance. Typical sources are self, family, directors' loans and retained profits, debtors, stock reduction and extension of credit from suppliers, trusts, clearing and other banks (overdraft), merchant banks, finance houses, leasing companies and factoring and invoice discounting companies. Medium-term funding (2 to 10 years) is for medium-term assets, plant and machinery, hard-core working capital, research and development. Long-term funding is above all for land and buildings and possibly some other long-term assets. Sources of medium- and long-term funding include self, family, directors' loans and retained profits, clearing and other banks, some building societies, development corporations and boards, merchant banks, finance houses, leasing companies, central and local government loans and grants, EU loans, insurance companies and pension funds.

Business finance

Ordinary shares

An ordinary share is the unit of investment capital which an investor pays in when he sets up a company. The ordinary shareholders of a company are its owners. Each year the shareholders receive some of the profits of a business. The slice of the profits given to the shareholders is a dividend. For example, a share may be worth £20. The annual dividend may be 50 p or £2 or any value, depending on how much profit the directors decide should be retained in the business and how much should be distributed to shareholders. To raise additional capital a company may issue new shares. It sells these shares to the public, banks, pension funds, insurance companies and other financial institutions. The company can then use the money it has raised for new investment. Shares are traded on stock exchanges throughout the world. If people think that a company has good growth and profit prospects they will buy the shares and the value of the company will go up. Conversely, if a company has poor prospects the value of the shares will decline. Trading in shares is done independently of the company. The rise and fall of its share price does not alter the flow of cash in the company's normal operations. However, businesses often prefer a high rather than low share price. Apart from anything else a high price usually makes it easier for the company to raise more capital if it needs it because it is perceived as being a success.

Preference shares

Preference shares are more like loans than ordinary shares because the owners carry no voting rights in the company. They are called 'preference' because if the company goes bankrupt the preference shareholders are paid off before the ordinary shareholders. Preference shares are now an uncommon means of raising capital. In general, companies prefer to issue *debentures* or *bonds*.

Debentures

Debentures are also known as *loan stock* and as *bonds* in the United States. They are certificates issued by a company, each with a face value. If a company wishes to raise £1 million it sells debentures to the value of £1 million in exactly the same way that shares are sold, to the public, pension funds, insurance companies, etc. The certificate will usually have a redemption date on which the company promises to pay back the loan, as well as the interest payments.

Key issues for suppliers of finance

Architecture

Suppliers of finance look for relationships inside the company, to produce at the lowest possible unit costs, impressive links with suppliers, evidence of market orientation and focus and evidence of enthusiastic customers. Where relevant, they might look for evidence that the business is aware of the gains from cooperation, such as inside a cooperative or producers' organisation, and when it is more important to compete.

Reputation

Where relevant, suppliers of finance to a service industry, such as a fisheries consultancy, would look for a good track record, which is a sign of reputation, and therefore the business is likely to have repeat purchases. Or if the business is supplying fish farms with inputs financiers might expect to see some other evidence of reputation.

Innovation

Innovation is important as a means of maintaining a business in a competitive environment, so financiers would expect to see evidence of production using the most appropriate production techniques. For example, an office should use computers rather than typewriters to prepare correspondence and a fish farm would plan to use modern, but appropriate, technology.

Strategic assets

A strategic asset in the form of a patent or a licence is very helpful because it signals that even if the business is not the most efficient in the world, there is something there to be exploited, with careful management.

Summary

Planning

The business plan is a key document. It forces the established business or the new entrant to think through what he is going to spend on costs and how those costs are going to be recovered and contribute to profit.

The role of information

There is a good amount of information available to businessmen in the UK, although probably not many are aware of what can be accessed. Generally, in developed countries a search can reveal such information, but the developing world is a different matter. It really takes a certain amount of perseverance and some intuition to get to know what is going on.

Stages of business inception

Whether it is an entirely new business or a new project within an existing business, getting the business going follows a similar pattern: identification, preparation, implementation and monitoring (keeping on track).

Basic rules of bank lending

These are the basic principles which commercial banks apply when assessing applicants:

- An equal contribution to the capital from the businessman
- Full security on bank's lending
- Capacity to service interest charges
- Capacity to repay loan
- The borrower himself.

Principal sources of investment finance

Development agencies, self, family and friends, shareholders and directors, internal company sources (not applicable to start-ups), banks, specialist financial institutions, grant/loan awarding institutions, venture capitalists, mergers and acquisitions.

Second-tier finance

Hire and lease purchase, contract hire, sale and leaseback, factoring and invoice discounting.

The principle of matching finance

The principle of matching finance is that money raised by a business should be matched to the purpose to which it is applied.

APPENDIX: EXAMPLES OF BUSINESS PLANS

Readers are referred to Birley (1979) for an example of a fisheries business plan. In that case the product was *Crassostrea gigas*, the Pacific or Japanese oyster. Two abridged business plans are reproduced here. They are shortened to preserve confidentiality and the financial forecasts have been removed. In neither of the cases were the companies concerned adventurous, but that is not surprising since they enjoyed some stability in the turbulent commercial environment of eastern Europe.

SC Ropes and Twines SA

This business plan was prepared for SC Ropes and Twines SA in 1995 by Andrew Palfreman and Chris Curr of the University of Hull International Fisheries Institute (HIFI). The authors would like to acknowledge with grateful thanks the assistance received in its preparation by the Managing Director, and the officers of the Marketing and Sales Department of the company. For reasons of confidentiality some details have been omitted. The authors also wish to express their thanks to the National Resources Institute, for inviting HIFI to undertake this work and for giving permission for this abridged version to be reproduced here.

Executive summary

SC Ropes and Twines SA is a company based in Romania, specialising in (1) the manufacture of fishing nets and fishing gear and (2) the manufacture of cords, ropes and twine. The company also makes various related miscellaneous items, such as nets for sports, hammocks, vegetable sacks in netting, heavy-duty nets for the offshore oil industry, hunting nets, nets for aquaculture and nets for various other uses. Sales in 1994 were about 4.7 billion lei (US$2.6 million) and the company employs 440 people.

The business is consistently profitable. Profit before tax as a percentage of sales income is between 10 and 16% from 1992 to 1994 and is rising. The business has no borrowings and hence a gearing ratio of zero. This is a reflection of the cautious approach to expansion adopted by the management. It implies also that if satisfactory projects could be found the company could safely increase its level of borrowing. It can be concluded that the company is fully justified in seriously considering market expansion and technical innovation.

The business and its management

History and position to date

The known records of the business go back to 1927. In 1948 the company was nationalised and became the Fishing Gear Factory under the Ministry of Light Industry. Within this regime it was an autonomous enterprise under the direct control of the State. In 1973 it became part of the Ocean Fishing Company. However, this did not prevent the fishing gear factory from selling nets to coastal and inland fishermen working in the Black Sea, the Danube Delta and on various lakes and farms, as well as meeting the requirements of the ocean-going fleet. In the 1980s the fishing gear factory became involved in the delivery of baler twine for agricultural purposes.

In 1991 the factory became a commercial company (SA) wholly owned by the state. Under its Articles the company has very wide rights to manufacture and trade, including importing and exporting, in the chosen field of activity – fishing nets, cords, ropes and fishing operations. Indeed, if the company wishes to widen the scope of its activities there would be no significant legal obstacles. It is simply a matter of registering the new areas with the Ministry of Commerce and the payment of a nominal fee.

The company employs 440 people of whom 9 are senior managers, a further 38 are in other executive positions and the remaining 393 are employed in the manufacturing process. The company owns a total of $17\,000\,m^2$ of land, of which $8000\,m^2$ is occupied by various buildings; $700\,m^2$ is rented out to other users. The company does not yet have the documentary title to all of its land but expects to do so within the very near future.

The company has a registered trade mark.

The business mission

The mission of SC Ropes and Twines is the manufacture of fishing nets, fishing gear, ropes and twines to the highest possible technical specification and product quality. The company aims to continue to be the only significant supplier of these items in Romania, to be a major supplier in the Balkan region of Europe and to participate actively in the global market for these products. It will continue to operate profitably, innovating and undertaking new investment when deemed necessary, but always acting with the caution necessary for an important estab-lished business in a turbulent commercial environment.

Objectives

- The objectives of the company are: to establish a chain of retail outlets for the sale of its products throughout Romania. In the long term it anticipates a total of 40 such dealerships. In addition to providing a major new commercial channel the planned retail outlets will generate market information, thus enabling the company to obtain a more accurate picture of the market for its netting products, ropes and twines, especially within the agriculture industry. The business plan is based on the assumption that it will establish 20 new retail outlets over 2 years.

- to invest in new extruding equipment for the manufacture of polypropylene yarn. In this way it will have control over an important raw material input into its production process and will reduce some transport costs. The cost of this equipment is estimated by the company to be about US$100 000. The company expects operating costs to fall as a result of the saving in transport costs and the reduction in purchases of polypropylene yarn from other suppliers.

Management team

The senior management team comprises the Managing Director, Technical Director and Director of Finance.

Legal structure

SC Ropes and Twines is a 'Societatea pe Actiuni' (SA – a company denominated in shares). Legal responsibility for the company lies with the 'Consiliului de Administratie' (Council of Administration – equivalent to the board of directors). In general, professional advice is obtained from within the company rather than by using outside specialists.

The product

The company is highly experienced in the manufacture of two related product groups: (1) cords, ropes and twines; and (2) fishing gear. In addition, the company is able to manufacture various miscellaneous related products such as nets for sports, hammocks, vegetable bags, safety nets for the offshore oil industry, nets for the protection of ski slopes, and netting products for agriculture and aquaculture.

Readiness for market

The various products of the company leave the factory ready for use. When accessories are required, such as the attachments to fishing nets, these are purchased by SC Ropes and Twines and attached at the factory.

Degree of monopoly

SC Ropes and Twines is the only significant producer of ropes, cords, twines and fishing gear in Romania and is in a very strong marketing position throughout the Balkan region.

Guarantees

The ropes and twines are produced by the company to Romanian standards and in accordance with international standards (ISO 9000–9004). The company supplies a certificate with its products which guarantees the breaking strain of the ropes and twines.

Product development

The company has been alert to new markets. Its production of hammocks is an example of its ability to respond to a new market for consumer products. However, SC Ropes and Twines is essentially an industrial producer rather than a producer of consumer goods and expects its main markets to be in its traditional range of products.

Sources of supply

Most of the inputs required for the business can be obtained in Romania from Romanian manufacturers and are readily available. The only significant exception is sisal, which is imported into Romania, although not directly by SC Ropes and Twines.

Markets and competitors

Internal

The company's Romanian clients are now predominately fishermen operating in internal waters using light twine of 0.3 to 0.5 mm. The heavier twines of 1 to 2-mm diameter are more suitable for nets for marine fisheries, but the decline of the ocean-going Romanian fleet has meant that there is no longer a large market for the heavier nets. The company believes that the internal market for its products is substantial, but demand remains unsatisfied because the distribution system is ineffective. If the ropes and twines are offered more widely to potential users, especially farmers, the company believes that much greater sales can be achieved. As a result, the expansion of the distribution system is a major part of the company's marketing strategy and therefore its business plan.

External

Export sales have been somewhat disappointing for the company. Sales have been made to buyers in Bulgaria, Greece, Egypt and Hungary, but 1994 export sales are not expected to be much more than half the level achieved in 1993 when export sales were valued at about US$100 000. The company recognises that its commitment to expanding into export markets has, as yet, been less than wholehearted. One reason for this is that the company has only recently been involved as the principal in international trade, trading on its own account. Under the state-trading system, international trade took place through an agency. So its first-hand knowledge of international markets is at a relatively early stage of development. Secondly, its main market has been Romanian buyers so its concentration on the domestic market is supported by the fact that the returns are still substantially greater there than abroad.

Market segments

The company has identified the following market segments:

- Nets
  - the fishing industry in internal waters
  - the marine fishing industry
  - miscellaneous users.
- Ropes and twine
  - shipping and fishing industries (mooring ropes, flags, gear attachments, etc.)
  - agricultural users
  - sports
  - miscellaneous users.

Customer decision criteria

Fishing gear: The demand for fishing gear is determined wholly by the current requirements of the industry. Fishermen are able to go to gear suppliers and specify their requirements with some precision. The gear can then be made up to suit them. Some examples of factors which might influence the fisherman's decision are as follows:

(1) The fishing method employed. A pelagic purse seine fishery will require relatively flexible nets that do not break under the heavy strain of large volumes of fish. The company now sells many nets to small-scale inland fishermen. They require gill-nets which are as invisible as possible to the fish. Twine made from polyamides is likely to be the most suitable material.

A trawl fishery for demersal species for human consumption may require a great deal of toughness in the gear used, especially if operations take place over rough ground, but elasticity may be less important. Bottom trawling typically requires the following characteristics:

- excellent abrasion resistance
- does not hold sand
- floats
- good wet knot strength
- good knot stability
- low cost.

These requirements are met by braided polyethaline, with a preference for green. Nets for mid-water trawling need to be strong so polyamides may be selected. Mid-water trawling requires the following characteristics:

- low drag
- high tenacity
- good wet knot strength
- good wet breaking strength.

The more elastic polyamide multifilament twine meets most of these requirements.

(2) Legislation governing a fishery. Even if netting meets regulations initially, the fishermen will wish to be sure that it remains within the rules over time. Various factors, such as sand and breakages, can cause changes in mesh sizes.

(3) Ease of repair, durability and colour. Some materials are easier to mend than others. Some materials deteriorate faster than others. The colour of the twine influences the ability of the net to capture fish in different conditions.

(4) The salesmen for the gear. If a fishermen can be persuaded that he loses nothing in catching capability by opting for a product sold at a lower price, he may well be willing to try a new supplier. If the gear then fails to deliver, the reputation of the manufacturer will be damaged.

(5) Price, but not to the exclusion of other factors.

The industrial users of ropes and twines purchase according to their specific requirements. Price is important, but considerations of the match of the product to the specific requirements of the customer are fundamental.

Competition

As noted above, the company is the only significant supplier of fishing gear and ropes in Romania. However, it is aware that the increasing liberalisation of the economy could change this. It may become possible for other businesses to source their gear requirements in other countries and thus internal competition will intensify. Externally the manufacture of fishing gear, ropes and twines is highly price-competitive.

Competitive business strategy

There are four main elements in the competitive business strategy of SC Ropes and Twines:

(1) The company will consolidate its existing production and selling system. The company works well, it is able to operate flexibly to meet customer

requirements and operates at a profit. A steady improvement in its technical capabilities is envisaged. It will gradually buy new machinery and develop new products.

(2)   The company is convinced that there are significant and rapid returns to be made from expanding its activities in the Romanian domestic market. One of its principal competitive developments is, therefore, to expand the number of shops which retail its goods. The company plans to cover the whole country with approximately 40 outlets eventually. The business plan has been calculated on the assumption of 20 outlets.

(3)   The forward integration of the business into new marketing activities will be complemented by upstream integration into the production of raw materials. The company intends to invest in the machinery for extruding its own polypropylene. The company believes that this will result in a saving of operating costs as it will be able to obtain its own requirements for raw material more cheaply as well as save the cost of transport from current suppliers.

(4)   The company recognises the importance of export marketing and plans to devote more resources to this area of activity, although the precise plans for this are not yet formulated.

Selling

The company sells most of its products directly, but it has recently adopted the practice of appointing representatives in various locations who sell the goods on commission. They take orders and the goods are then delivered within 3 or 4 days. The representative arranges the sale documents and remits the cash receipts to the company. All goods are sold with a certificate guaranteeing its technical characteristics. The company does not engage in credit sales. In general, it asks for 10% of the agreed price with the order and then the remainder of the payment is made on delivery, although other arrangements can be negotiated. Few orders take longer than a month to prepare and most can be ready within a few days.

Manufacturing

Production process

The staff and equipment of the business can handle a wide variety of materials and can respond very quickly to the requirements of customers. The list includes nets made from a variety of materials as required by customers, polyamide, polyethaline and polypropylene cords and ropes, multifilament nylon twine and baler twine. The company can also manufacture with single filament twine and various products from natural fibres.

The production process for nets is as follows:

(1) The preparation of yarns by twisting and winding onto bobbins
(2) The manufacture of nets
(3) Heat treatment of the nets with steam
(4) Drying the nets
(5) Dyeing (if required)
(6) Finishing and assembly of nets using floats, weights and other extras from Romanian suppliers
(7) Checking, control and inspection.

Production trends

In the period 1988 to 1991 total production from the factory has declined from 6000 to 800 tonnes. However, this decline is largely accounted for by the fall in the output of cotton products from 370 to 85 tonnes per year, and a fall in the production of baler twine from 4400 tonnes in 1988 to 40 tonnes in 1991. The decline in the production of cotton products followed from the decline in carpet manufacture in Romania. The high levels of demand for baler twine in the 1980s followed from government regulations requiring farmers to avoid using wire. Agricultural methods today no longer require the same amount of twine although some improvement in the market seems to be indicated.

The machinery and other equipment owned by the company are summarised in Table A 7.1.

Table A 7.1 Summary of current machinery and vehicles.

Description	Number	Origin	Year of manufacture
Winding machines – twines	40	Japan, Romania, Germany	1948–1979
Twisting ropes and cords	30	Italy, Germany, Japan	1980–1989
Winding machines – ropes	55	Romania, Italy	1980–1984
Heavy net making	7	Germany, UK	1948–1974
Other net making	14	Japan	1973–1987
Stretching, heating	4	Various	1965
Three trucks, two tractors, forklifts	na	Romania	Various
Workshop tools	na	Romania	Various

na, Not available.

Machinery requirements

- The company finds that it is sometimes working at the limits of its capacity in the production of fine twines for the production of nets for internal waters.
- The company believes that there is a market for ropes of a larger diameter than currently manufactured and offered for sale and therefore may purchase a rope braiding machine for this.
- The company intends to purchase the equipment for extruding polypropylene yarn.

Finance

Table A 7.2 is a summary of indicators from the company's balance sheet in 1992 and 1993 with some provisional ratios for 1994. The key points to note about the balance sheet are:

- The business is consistently profitable. Profit before tax as a percentage of sales income is a preferable measure than the return on investment because the latter is based on non-market definitions of the capital employed in the business. Pre-tax profit as a percentage of sales lies between 10 and 16% of sales in the 3 years under examination and is rising.
- The business has no borrowings and hence a gearing ratio of zero. This is a reflection of the cautious approach to expansion adopted by the management. It implies also that if satisfactory projects could be found the company could safely increase its level of borrowing. An assumption that some of the new investment will be financed by borrowing is included in the business plan.

The business plan

(Detailed calculations and tables omitted.)
The discounted cash flow analysis shows that the internal rate of return to the proposed investments in new retail outlets and new equipment is high at 53%. The

Table A 7.2   Balance sheet indicators.

	1992	1993	1994
Gearing[a]	0	0	0
Liquidity[b]	10.4	3.2	na
Return on investment (%)	4.6	21.3	40 (est.)
Profit before divided by tax sales (%)	10.3	15.5	16 (est.)

na, Not available.
[a] Ratio of debt to equity; the business has no loans, only share capital.
[b] Measured by the current ratio, or ratio of current assets to current liabilities, where 'current' means payable within 1 year.

internal rate of return and the net present value of the new investments are very sensitive to changes in sales revenue and changes in direct and indirect costs. For example, a mere 5% reduction in forecast sales revenue reduces the internal rate of return to close to zero Similarly, a 5% increase in direct and indirect costs would reduce the internal rate of return from over 50% to under 20%.

The balance sheet forecast shows that if the proposed new investments make the predicted contribution, the net worth of the business increases dramatically (by a factor of 3). This takes place with little impact on the debt to equity and gearing ratios.

The overall conclusion is that SC Ropes and Twines is absolutely correct to seek additional profitable business opportunities. At present it is a successful business and how expansion might be financed remains an open question.

Fish Farm Society

Introduction

This summary business plan was prepared for the Fish Farm Society in 1995 by Andrew Palfreman and Chris Curr of the University of Hull International Fisheries Institute. The authors would like to acknowledge with grateful thanks the assistance received in its preparation by the Directors of the Company (named in original). The author is also grateful to the National Resources Institute for permission to reproduce this summary of the original. The authors of the plan were aware that they were reflecting the extremely cautious approach of the business managers.

Executive summary

Fish Farm Society was established as a state-owned fish farming business in Romania in 1958. At the outset it was responsible for the management of fish farms with an area of 650 ha. In 1991 it became a company in which the government was the sole shareholder (SA). By 1991 the land and water area managed by the company had grown. It manages a total of 12 fish farms and expects production of final output in 1995 to be about 2100 tonnes, worth about 2.1 billion lei (about US$1.2 million).

The company has three fundamental problems and corresponding strategies for dealing with them:

(1) Its current production is dominated by Chinese carps (about 90%). These are not seen as popular with Romanian customers as common carp. The strategy of the company is, therefore, to shift production from Chinese carps to common carp.
(2) The company does not have legal title to all the farms it is currently managing. It is taking steps at a local and national level to secure legal title.

(3)   The company is plagued by poaching. This is because many people living in rural areas perceive fish from the farms to be public property. The company recognises that this is an extremely difficult problem to deal with, but through improved protection of land and seeking the cooperation of the army it hopes to achieve more security.

The company is financially stable, albeit not very profitable. The business plan prepared for the Company forecasts a similar prevailing pattern of relatively low levels of profitability, but financial stability. However, to achieve the market expansion planned for and required if stability is to be achieved, some new investment is required. This will consist of some working capital and investment in six shops.

The key financial problem for the company is that its cash flow is very sensitive to sales revenue. Since it operates on very tight margins, failure to achieve the forecast levels of sales could have serious consequences for the cash flow.

The business and its management

History and position to date

The company was established as a state-owned fish farming business in 1958. At the outset it was responsible for the management of fish farms with an area of 650 ha. In 1990 it became a company in which the government was the sole shareholder (SA). By this time it had expanded its surface area to 3500 ha, of which 1500 ha consisted of artificial reservoirs.

The company consists of 12 fish farms (named in original). A number of the farms consist of a mixture of dams and ponds. Some of the ponds are fed by natural water flow whilst others receive pumped water. Until June 1991 the company was engaged in the production of processed fish but abandoned this because of a lack of markets for the output. The company sells most of its output through its four shops, and some to other private and state outlets. It also sells to fish canners from time to time but regards these as distress sales because the price offered is so low. The production of fish by Fish Farm Society is shown in Table A 7.3.

Table A 7.3   Production of fish by the company.

Year	Tonnes (predominantly grass carp)
1991	1900
1992	2000
1993	1000
1994	2000 (est.)

The company is not a very profitable business at present, but it is meeting its costs and making a small contribution to shareholders' funds in each year. Its profits were between 3 and 4% of costs in 1993. It is important to add that profit maximisation as such has not hitherto been a part of its business strategy since this would lead to higher tax payments by the business and could also imply lower levels of employment. The business employs 200 people of whom about 10% have university degrees.

The company faces two legal problems: (1) It requires legal title to the fish farms which are under its management. It is without legal title to approximately 650 ha. (2) The need to adjust supply to demand under conditions of fluctuating markets is not helped by the legal obligation to give 30 days' notice of price changes. This is particularly problematic in the face of competition from private sector retailers.

The business mission

The mission is to produce good-quality farmed fish for the Romanian domestic market. It will achieve this while continuing to make a profit. The company expects to make incremental improvements by shifting the balance of fish produced towards more marketable species, notably common carp.

Objectives

The objectives of the company are:

- to shift the balance of production away from Chinese carps towards common carp and caras (*Carassius carassius*);
- to reduce losses of fish through poaching;
- to achieve legal title to the fish farms under the management of the company.

Management team

The *Consiliului de Administratie* (the board of directors) is directly responsible to the shareholders and is appointed by the state. It consists of the Director General, Commercial Director and Finance Director. The management team is an experienced and highly competent group. Its members work well together and have a very clear shared vision of the future of the company. They have no outside professional advisers but rely on the technical capabilities of themselves and other company employees for advice. The Consiliului has a contract with the State Ownership Fund and the Private Ownership Fund for the management of the business.

Legal structure

The company is a *Societatea pe Actiuni* (SA) and was registered in 1990. All the shares are owned by the state. Seventy percent of the equity is owned by the State Ownership Fund and 30% by the Private Ownership Fund.

The product

The Company produces whole fish. Their current price range is between 780 lei per kilo and 2130 lei per kilo depending on species and size. As a matter of policy it has rejected the production of processed or cut fish because it does not believe that it is profitable to supply them to the market. The company has also rejected selling gutted fish because it takes the view that consumers prefer whole (ungutted) fish, which is perceived to be fresher.

Comparison with the competition

There are two competitors within the region. However, both of them are smaller than Fish Farm Society and can only produce about 100 tonnes of fish per year between them because of limited productive capacity. Meat products are in competition with fish, but in general these are significantly more expensive, approximately double the price of fish.

The market and competitors

The market for the farmed species in Romania is no longer clearly defined. Before 1989 total fish supplies were perhaps in the order of 40 000 tonnes of inland fish and a similar quantity of marine species. However, the supply of marine species is now very much reduced so the total availability of fish is much reduced, although to levels about which there are no hard data. Where once Romanian people in general were pleased to absorb the output of the fish farms, lower real incomes and a wider variety of goods on offer have conspired to reduce the demand. Many of the outlets which used to accept fish for sale have switched to general food distribution.

Nevertheless, there is still a requirement for fish and this is being proved in practice every day by the active selling practices of businesses such as Fish Farm Society. But where a new equilibrium will be reached between consumer demand and supply by fish farms remains to be seen. Fish to the Romanian domestic market is supplied under competitive conditions. The company now has to fight for markets in competition with other fish producers and other foodstuffs.

Customers

In general, the buyers of fish are of mixed ages and genders. However, it is possible to judge that they are more often than not people on relatively low incomes. This might imply an income of 150 000 lei per month or less. The appeal of fish to the lower income consumers is its lower price at a time of declining real incomes.

The company is not pessimistic about the future of the market. It believes that it can look forward to growth in the future when real incomes begin to recover combined with its own attempts to improve the market for fish.

Competitive business strategy

(1) The company intends to change the composition of output from the current pattern of predominantly Chinese carps (90%) to the production of common carp and caras. It is aware that the market for Chinese carp has declined, for example the volume of Chinese carp sales in the main city of the region has fallen from 10–15 tonnes per day before 1989 to 3–4 tonnes in 1994. It expects market demand for common carp to be stronger than for Chinese carp and therefore output from the business can be increased. The planned production is listed in Table A 7.4.

Table A 7.4 Planned fish production (tonnes).

Year	Common carp	Caras	Chinese carp < 1 kg	Chinese carp > 1 kg
1994	230	150	1500	350
1995	450	250	500	550
1996	550	350	500	200
1997	750	350	400	500
1998	750	350	300	500

(2) The company plans to promote the sales of fish. It has an operating budget fixed at around 2% of operating profits for fish promotion and has experience of improving sales through local television and radio advertising and the supply of recipe leaflets. However, crucial to its plans is the opening of six more shops in addition to the four already operating.
(3) The company plans to offer fish for sale more actively throughout the year. The shift in production to common carp and caras will facilitate this because their shelf life is better than for Chinese carp in the warmer weather.
(4) Obtaining title to the fish farms is fundamental to its long-term plans. It is taking the necessary legal action to obtain the required rights.
(5) A final element in the company's long-term plans is to reduce poaching. This is not easy under the circumstances of the business, with properties spread out over a wide area. It also appears to be traditional practice for country people to obtain their fish requirements through taking it from the farms. However, through negotiations with the army it is hoped that some improvements can be obtained. The company is also working independently and through its trade association to secure larger fines for poaching and the implementation of a law on fisheries in accordance with the new conditions.

Selling

Fish is sold either by the company or by the fish farms themselves. About 30% of production is sold through the four shops owned by the company. A further 30% of production is sold to private shops and other traders. The remainder is sold to state-owned shops. The speed of payments for fish is not thought of as a problem by the company.

Selling to fish canneries has been largely abandoned because the price currently offered – about 900 lei per kg – is not remunerative. There is a further problem in selling fish to canneries, the frequent delay in payment – from 6 months to a year. Delaying payment beyond 30 days is in fact illegal in Romania, but the problem is widespread.

Production

The fish farms divide into three main categories.

(1) *Closed systems of production.* These farms are closed systems with control over their own water supply. In some cases the water is pumped into ponds whereas in others it is possible to use gravity and water from reservoirs.
(2) *Open systems of production.* This production takes place in natural earth ponds. In this the flow of water cannot be controlled as it depends on river supplies and rainfall.
(3) *Fish production in reservoirs.* These are stocked with brood stocks and fished each year.

Each farm within the company has its own production plan determined by the headquarters. Each farm obtains its fingerling requirements from within the company. This avoids the need to transport small fish and therefore reduces risks. The company believes it is well equipped with the capital equipment and land to increase production and therefore requires no significant new investment in the farms. The company believes it could produce about 4500 tonnes from existing facilities without further effort. In fact, total fish production in 1994 will be about 3000 tonnes of which about 2000 tonnes will be marketed. The business has pumping stations, gear for excavating and a number of vehicles. For example, it has three 7-tonne insulated trucks, two 1-tonne insulated trucks and a 16-tonne vehicle which is no longer in use. It has various tractors and other mechanical equipment for the farms and four insulated trailers.

The company has about 200 employees in total, mostly spread among the various fish farms. The business is considering a reduction in the number of permanent employees and their replacement by part-time staff. However, it is cautious about taking this step because of the requirement for well-qualified and experienced teams for operating the fish farms. At the head office are the following staff:

Accounts: 5
Production and investments: 3
Selling, transport and supplies: 3
Administrative services (including a legal adviser):   3

Finance

The balance sheet for 1994 and the previous 2 years is summarised in Table A 7.5. The four main conclusions which can be drawn from this are:

(1) The company has successfully weathered difficult trading conditions for the last 3 years. Nevertheless, it has met its operating costs.
(2) The gearing ratio (all loans, including overdraft, divided by capital employed, including overdraft), a measure of debt to equity, is very low in each of the 3 years. This is because borrowings relative to equity in the form of Social Capital are consistently very low. It is a good demonstration of the very responsible management of the company. If the company were interested in major expansion this ratio of borrowed money to equity could be allowed to rise to finance new projects.

Table A 7.5 Summary balance sheet (in '000 lei).

	1992	1993	1994 (est.)
Fixed assets			
Cost	1 949 324	1 956 967	14 088 937
Depreciation	102 297	181 874	260 874
Net book value	1 847 027	1 775 093	13 828 063
Current assets			
Stocks	81 044	234 209	891 400
Debtors	31 298	32 187	25 000
Cash at bank	2 365	4 134	0
Total current assets	114 707	270 530	916 400
Current liabilities			
Creditors	5 641	8 073	70 000
Other debts	93 460	79 178	430 000
Taxes	10 059	27 167	90 000
Total current liabilities	109 160	114 418	590 000
Adjustments	27 972	20 314	90 000
Net assets	1 880 546	1 951 519	14 244 463
Financed by			
Shares/owners' capital	1 857 785	1 851 219	13 813 652
Accumulated profits	2 172	27 718	75 000
Loans	20 589	72 582	355 811
Total	1 880 546	1 951 519	14 244 463

(3) The liquidity ratio at greater than 1 in each of the 3 years is more than satisfactory since it shows that the company would be able to meet all its short-term debts from the sale of its liquid assets in the event of an emergency.

(4) Profits expressed as a percentage of ownership interest were low in each of the three years.

A summary of performance is given in Table A7.6.

Table A 7.6   Summary of performance indicators.

	1992	1993	1994 (est.)
Sales (million lei)	105 000	336 000	806 000
Direct costs divided by sales	78%	73%	56%
Net profit before dividends and tax divided by sales	4%	0%	3%
Gearing[a]	1	6	5
Liquidity [b]	1	2.5	1.5

[a] Ratio of debt to equity.
[b] Measured by current assets divided by current liabilities, or current ratio

Business controls

The company operates as a standard, well-established Romanian business with all the normal business controls in place. That is, there is an established system of controls over payments and receipts, fish farm activities, sales and personnel. In the shops staff are required to balance recorded sales and cash receipts each day and there are periodic checks on the deliveries, stocks and fish sales.

Conclusions of the business plan

• The business plan includes the purchase of six new shops. Apart from this its main element is a shift in production from grass carp to other species.
• The business plan follows on the current performance of the business. However, to achieve the marketing push intended by the company the analysis suggests a requirement for working capital. The base case suggests a requirement for 900 million lei if resort to overdraft financing is to be avoided. This follows from the narrow cash margin on which the company has operated and intends to continue.
• The gearing ratio continues at very low levels throughout the forecast suggesting that if profitable projects could be found the company should not be restrained by fears of excessive levels of borrowing.

- The net worth of the business, at about 14 million lei in 1994, is expected to increase only gradually, again following from the cautious strategy of the company.
- The analysis assumes some capital investment. The internal rate of return of the business given the capital investment the business intends to undertake (working capital and four shops) is high at 60%, as it should be bearing in mind that a great deal of capital already exists in the business.
- Both the internal rate of return and the net present value are highly sensitive to revenue changes. Therefore if the sales plan is not fulfilled for some reason the consequences may be quite serious. On the other hand, if the company can improve upon its predicted performance, the rate of return, and hence the ability to service debt, will be substantially greater.

Chapter 8
Aspects of Marketing for Fish Businesses

Marketing is about putting distinctive capabilities (see Chapter 4) into an acceptable form and presenting them to selected market segments. What is really difficult about marketing in the fish industry is actually understanding what the market requires. The characteristics of fish products can be very subtle and it requires real knowledge and experience to be aware of these. See, for example, Justin Webster's informative article about tuna, 'The Big Blue', in *The Independent Magazine* (Webster, 1998), for an assessment of how exacting the Japanese market for sashimi-grade tuna can be and how difficult it is for non-specialists. Consider also the various categories of shrimp which are traded internationally, noted in Chapter 9. Garuccio (1995) prepared a selective bibliography on fish marketing for FAO. Through many publications and circulars, and especially through the Infofish umbrella, FAO has maintained an excellent worldwide service of fish market information. Infofish is the network of regional fish marketing and information services established by the Fishery Industries Division of FAO.

The complexity of the market is well-illustrated in the edited extract from the Yorkshire and Humberside study (Palfreman *et al.*, 1993) reproduced on pages 109–114. A variety of different types of business are involved in the fisheries sector. There are those who produce fish, and this category subdivides into fishing vessels and fish farms. There are others who process fish. Processing is often split into two categories: primary processing, referring essentially to the basic treatment of fish such as gutting and filleting, and secondary processing, referring to the conversion of primary processed fish into consumer products. Some businesses transport fish and others supply the inputs for the fish industry, such as vessels, engines, excavation of fish farms and many other inputs. Fish retailing to the consumer is a major component of the industry and an important part of the trade is the sale of cooked fish to the public, from fast food outlets as well as restaurants. An increasingly important part of the industry is the international trade in fish, with specialists springing up who know sources and suppliers and are able to bring the two together, for a commission. So it is all very big and complex, and analysing the market is one of the keys to success.

Chapter 4 refers to architecture, reputation, innovation and strategic assets, brought to market, as the means of achieving competitive advantage and economic added value. Niche marketing – finding a special relationship with a limited number of customers – falls under the heading of architecture. This is the realistic place for any small business, especially new ones, because they cannot compete with the big bulk suppliers. The boxed extract below describes the fish processing and distribution system in one region of the UK. The ways fish is distributed throughout the world are many and varied.

The fish processing industry in the Yorkshire and Humber region: a case study

Box 8.1 gives an extract from Palfreman *et al.* (1993) *The Fishing Industry of Yorkshire and Humberside: A Regional Study* – with acknowledgements to William Horner who was mainly responsible for this section. The *architecture* of the Humber region's fish industry – the combination of skills, the business contacts, the fishing tradition and the training facilities – and *innovation* in modern fish processing technology, contribute to the region's continued success.

Box 8.1

Demand factors

The national and European demand for fish is always in a process of rapid change. Recent developments include the fresh fish counters in many supermarkets, increasing numbers of 'value added' products, and new forms of packaging and the export trade in dogfish. Input prices have put increasing pressure on the traditional fish and chip shop and small-scale fish retailing (fish mongering) continues to decline.

Primary and secondary fish processing

A distinction is often drawn between primary and secondary fish processing. Primary fish processing consists of gutting, filleting, freezing and other basic fish preparation. Secondary fish processing is the preparation of 'value added' food products from fish. Many companies in the region are engaged in both types of activity.

Primary fish processing in the Yorkshire and Humber region

Primary fish processing developed in the Yorkshire and Humber region precisely because of the very large-scale landings of wet fish over many years. A large supply of wet fish coming over the quayside is fertile ground for the rapid development of large and small fish wholesalers. There is a real economic linkage between wet fish landings and primary fish processing.

   Today there are a number of companies, large and small, who are engaged in primary fish processing. These include virtually all the members of the Grimsby Fish

Continued

Box 8.1 continued

Merchants' Association and of the Hull Fish Merchants' Protection Association (their trade association). There are also other direct fish distributors who are not members of the trade associations and there are many small-scale fish merchants operating in the small ports of the region. The lifeblood of these companies is whole fish supplies. Their incomes are derived from turning whole fish into products which are acceptable to consumers or to the secondary processors.

Source of supply of wet fish

There are basically three current sources of wet fish supply to the primary processing industry in the Yorkshire and Humber region: (1) landings by regional vessels; (2) overland fish, such as fish from the Yorkshire coast and Scotland; and (3) various other sources of supply: foreign landings, container fish and thawed fish from Iceland and the former Soviet Union. The supplies from non-regional vessels and other sources are attracted to the Humber because of the concentration of processing capacity and fish handling skills which were built up in an era of greater local landings, but now they form perhaps around 80% of the supplies going to the local industry.

The significance of variety in fish supplies

The fish processing business comprises a complex array of operators many of whom are continuing to develop markets and products. One well-known Grimsby company has expanded the market for dogfish and handles about 10 000 tonnes a year. A large Hull concern has specialised in the production of blocks of fillets. The largest fish processor in Grimsby has retained a highly flexible hand-filleting line at the same time as building up a substantial business in value added products. Small-scale operators in Scarborough and Whitby have established their trade on the strength of locally available supplies. When looking at the fish processing industry it is essential to recall that it is not a homogeneous industry. Rather it is an industry where there are many different niches. A heterogeneous supply of fish, especially wet fish, is the source of the Yorkshire and Humber region's diversified and successful fish merchanting industry. Thus, volume is important to processors but for the optimal development of the industry, so also is variety. This point deserves emphasis because one of the strategies adopted by the region's fishermen is to move towards species which are not yet subject to controls. All the three Fish Producers' Organisations in the region include a very wide range of target species in their fishing plans. One example of this type of diversification is the rapid expansion of the crab fishery off the Yorkshire coast.

Continued

Box 8.1 continued

Secondary fish processing

The secondary fish processing sector in the region is very adaptable and dynamic. Businesses have been adaptable and have responded well to changing markets. The demand for fish by secondary fish processors makes little distinction between fish purchased from local fish processors and imported blocks of fillets. Their main concern is that the specifications of the product in terms of species, quality and size should be exact and that the price should be right.

Some large companies (such as Birds Eye Wall's) have now left primary fish processing altogether to concentrate on secondary fish processing – the preparation of products which are profitable in an expanding market. Birds Eye Wall's is part of a multinational, Unilever, and they see primary processing as a cost that they can escape. Their expertise is more profitably applied to efficient production of value added products and marketing. Nevertheless, this company still obtains some of its raw material from regional sources. Other companies who are involved in the preparation of value added products, such as Bluecrest, continue to value sourcing their supplies from regional wet fish markets.

Fish processing

Long centred upon the deepwater ports of Hull and Grimsby, the fish processing industry provides employment for around 7600 individuals along the entire coastal zone of the region. In spite of the large decline in fish landings since the mid 1970s at these major ports, fresh fish supplies, sourced from the remaining regional vessels, the Yorkshire coast ports, Scotland, Iceland and Continental Europe, and frozen fish from all parts of the world, have ensured the industry's continued strength in the region. Two major factors sustaining this strength and the prospect for further investment are the location in Humberside of the highest concentration of cold storage space in Europe (0.7 million m^3, in addition to that belonging to individual food processors, in Hull and Grimsby alone) and a workforce well acquainted with the health requirements of food handling.

The role of cod

Cod, the cheap protein mainstay of institutional feeding until the late 1950s, became the most familiar of convenience foods in the form of the frozen fish finger in the 1960s and still remains the overwhelmingly predominant component of a wide range of value added frozen fish products. This dependence of the British consumer on this single species, together with the loss of access to distant-water fishing grounds, has resulted in a global pursuit of supply sources by the region's processors and some careful diversification away from cod towards other species, such as pollack and Pacific cod. About 25% of wet fish for primary processing and about 10% of raw material for the frozen fish product market is now sourced from within the region.

Continued

Box 8.1 continued

The consumption of fish compared to meat

Consumption of fishery produce in the UK varies between one-sixth (2-adult households) and one-ninth (2-adult + 4-or-more-children households) of the consumption of meat and poultry products, yet such has been the popularity of the fish finger and its fish block-derived successors that frozen fish products account for 25% of the total frozen food market. Reflecting the increase in white fish prices against the fall in poultry and pig-meat prices, people spend between one-quarter (2-adult households) and one-sixth (2-adult + 4-or-more-children households) of the amount they spend on meat and poultry products on fish and fish products.

The industry's labour force

The industry has always been one with a high proportion of female labour predominantly in semi-skilled or unskilled low-earning jobs. This situation is now changing with a larger proportion of women taking higher earnings in managerial and skilled (e.g. filleting) posts. The numbers in employment in the region's fish processing industry have declined significantly as have the number of small, primary processing firms over the last 20 years. Rather than being an indication of a general decline in the importance of the industry here, however, the reduced workforce reflects the developments towards automation on production lines. At the same time, there has been a concentration of activities amongst those enterprises prepared and able to bring their processing facilities into line with the EC Directive 91/493/EEC governing the health conditions for the production and placing on the market of fishery products.

The white fish frozen block and product development

The frozen 7.5-kg white fish block, from being just an intermediate in the production of fish fingers, has become an internationally traded commodity. Profit margins have been tight and this has encouraged product development to secure more added value. Negative consumer attitudes towards food additives, provoked during the early 1980s, resulted in a flurry of developmental activity. Diversification of the product range helped some block manufacturers through less profitable periods, but only two British firms (one in Humberside and one in Aberdeen) remain in the business today.

Technological progress

The major British value added fish products are frozen. To meet the demand in this market, the processors of the region have made very substantial investments in state-of-the-art process line equipment. Filleting and skinning machines, flesh-bone separators (to maximise the yield of flesh), belt-freezers, spiral-freezers, regulation-

Continued

Box 8.1 continued

reforming presses, continuous battering/breading and deep-fat frying equipment have, in most cases, become essential items in securing operational profitability. Although machine-filleting is fast, hand-filleting is still preferred by the majority of processors for its flexibility on fish size and quality, the better yield and fillet quality attainable and the easier control of process hygiene. Filleting machines have become very sophisticated but still need considerable design improvement.

Quality management

The top supermarket companies (Sainsbury, Tesco, Argyll, Asda, Isosceles, Marks & Spencer, Kwiksave and the Co-op) have become overwhelmingly influential with a total share of over 75% of the retail food market, as have the big fast-food operators in the catering sector. Fish processors are working closely with these major customers to comply with ever more rigorous product safety and quality specifications. The British Government's Food (Safety) Act 1990 introduced, for the first time, the concept of 'due diligence' to all those involved in the sequence of operations to the point of sale. Its effect is to switch the burden of proof in litigation from the prosecution to the defence over the wholesomeness or quality of a food item offered or intended for sale. This has triggered a profound shift in the approach to management of the food manufacturing and marketing industry. 'Quality' had been 'controlled' by periodically checking selected properties of a small sample of the end-products of certain unit operations in the manufacturing sequence. This had been long recognised as ineffective and progressive manufacturers had moved, or were in the process of moving, towards a regime of 'quality assurance' by 1990. Now manufacturers are working closely with their raw material suppliers and retailer-customers to ensure that wholesomeness and quality of the material being processed controls the selection of operations, operating conditions, equipment and personnel. Getting the appropriate quality in each individual item bought is seen to be everyone's right; therefore a product's quality should be part of its design: the responsibility of all involved in the harvesting/handling/processing/marketing sequence. [A summary guide to the Hazard Analysis Critical Control Point (HACCP) technique is included as an Appendix to this chapter.]

Quality has been the motivating factor in Humberside processors' recent move away from Icelandic sourcing of white fish (because of experience with high parasite levels) and towards sourcing frozen-at-sea and as much 'fresh' as possible from the Yorkshire coast ports' inshore fleets. The latter can reliably give the supermarket companies the length of in-store shelf-life they require for their increasing wet fish trade. Their share of this has increased to 29% of the total wet fish market since the introduction of supermarket fish counters 10 years ago. Supermarkets without wet fish counters, however, kept up with the perceived consumer trend towards 'fresh-rather-than-processed' fish by moving into the market for chilled 'modified atmosphere packaged' (MAP) fish. Again, principally Humberside firms were responsible for making the developments and capital investments necessary to supply this new market.

Continued

Box 8.1 continued

Recent developments

In spite of the number of small processors leaving the industry recently, the concentration of processing facilities in the hands of manufacturers who can quickly and hygienically handle large volumes of cheap fish suddenly coming onto the market is an advantage that Yorkshire and Humberside ports have over competitor North Sea fishing ports. Part of this flexibility stems from the diversity within the local industry. Besides freezing fish, the region's processors salt and dry (95% for the export market), marinate (for the delicatessen market) and smoke, often using species and sizes of fish unpopular in the fresh and frozen trade. They also supply the markets of the Far East with shark's fin, the German market with dogfish bellies for 'schillerlocken', the largest cannery in Europe (Pedigree Petfoods, Melton Mowbray), plus the region's fishmeal factory and fish oil refinery with the cheap by-products which are their starting materials.

Conclusion

The Yorkshire and Humberside fish industry is at the forefront of technological developments in fish catching and fish processing. On the catching side, vessel owners have been willing to diversify into new technologies to take advantage of developing opportunities. The growth of gill and trammel netting, and the development of a significant offshore crab fishery, are examples of this. In fish processing the region is at the centre of the European fish industry. It is in a strong strategic position because of its traditions, knowledge and state-of-the art technology.

Market analysis

The list of references at the end of the book includes the well-known text by Kotler (1997), *Marketing Management, Analysis, Planning and Control*, which is a comprehensive, clear and informative survey of general marketing. The two texts by Chisnall (1986 and 1991) are excellent guides to marketing research techniques and Crimp (1990) is also very useful. Curtis (1994) is the scientist's perspective. For the small businessman, Rosthorn *et al.*'s (1994) *The Small Business Action Kit* is a very good guide to marketing as well as many other aspects of small business management. Three of Rosthorn *et al.*'s checklists have been altered and included as Figs 8.1, 8.2 and 8.3 in this chapter.

Identifying a market segment

The marketing terminology for finding a niche is market segmentation. It means *identifying* the people or businesses who are likely to want your particular product. If the market segment can be identified the product can then be tailored to their particular needs.

Year	Total market for the fish product (kg or other units) (a)	Other supplier A	Other supplier B	Other supplier C	Own share (%) (b)	Units (kg or other units) (a) × (b)	Unit price	Comments
1995								
1996								
1997								
1998								
1999								

Fig. 8.1 Market shares and prices. (Adapted from Rosthorn *et al.*, 1994.)

Comparison with competitors	Worse	Same	Better	Comments
Product attributes				
Unit weight				
Packaging medium				
Packaging				
Presentation				
Appearance				
Delivery				
Colour/flavour/odour				
Image				
Species				
Freshness				
Source				
Quality of advertising				
Others				

Fig. 8.2   Comparison with the competition. (Adapted from Rosthorn *et al.*, 1994.)

In the catching sector the principal source of EAV (economic added value – see Chapter 1) is the efficiency of the catching operation. One vessel is more profitable than another because of skipper skill and technological advantages. In the market place the law of one price often prevails because the product is frequently homogeneous. However, this is not always so. Skippers can obtain advantages by targeting species which are more likely to fetch a better price in the auction. Also in recent years, with the advent of fish processed at sea, it is possible to improve the price received by careful fish processing on board. Or, to look at the same point in reverse, some vessels have earned themselves a poor reputation because processed fish quality has not matched the quality of other vessels.

The scope for identifying a market niche becomes more clear cut in the case of farmed products. For example, a fish farmer producing rare aquarium fish may be able to identify a few customers who are looking especially for particular species and will be willing to pay for the special fish. Similarly, the smaller fish processors may be able to identify a small number of restaurants who want to buy particular species of fish meeting specific quality standards. In this way the processor may

How many potential fish buyers are there?	
How many real customers are there?	
Who are they?	
Who does the buying?	
Where is the buying done?	
Where are the fish buyers from?	
What kind of fish products do they want to buy?	
Why do they want fish/fish products?	
Where do they get fish/fish products at present?	
How much will they want to pay?	
What deficiencies do current fish supplies have?	
When do customers buy, how much and how frequently?	
Can you deliver what they want when they want it?	
Who else can supply fish in this way?	
How strong is the competition?	
Will the market grow or contract?	
Will increasing supplies reduce the price?	

Fig. 8.3 Market analysis checklist. (Adapted from Rosthorn *et al.*, 1994.)

be able to compete with some of the larger companies. Retailers also may develop a relationship with a certain range of customers with particular requirements. Because they know their customers and are also expert in the fish business they can compete successfully against the larger companies.

Consumer market segments

When planning a fish business the entrepreneur should define, as far as possible, his target market. Typical categories of consumers are age, sex, location, job or occupation, special interests, lifestyles, attracted to certain events, seasonal purchasers, or any other important segment descriptions.

Other market segments

As is clear from the description of the Yorkshire and Humber scene given above, the larger fish processing companies are well integrated into large-scale high-tech systems of food processing and distribution. It is extremely difficult for small companies to break into this mainly because the investment required to produce fish products exactly matching the multiple retailers' requirements is so expensive. So, small businesses may have to look elsewhere or to collective action for their markets. Many small fish-processing businesses do not supply consumers. They have found a niche supplying fish and chip shops, the restaurant trade and specialist retailers.

Identifying customer needs

Once a suitable segment of the market is identified the next step is to find out what this group of people or businesses really need from the fish product. For example, quality may be very important, or a multiple may stress the importance of exact pack weights. Presentation, pack sizes, appearance, species, frequency of delivery, homogeneity and numerous other possible factors need to be identified. Asking people what they need is important in this context.

Market size

The analysis of markets now turns to the size of the market. It is, of course, of interest to major players to be aware of the total market size because then they can judge what their share is and by how much that share can be expanded. Smaller businesses are likely to be targeting a smaller market (such as fish utilised by fish and chip shops in the Yorkshire and Humber region). Obtaining that kind of statistic can sometimes be harder than global figures although, with ingenuity and a few questions to people, it is usually possible to reach a plausible estimate.

Market structure

Market structure is the system by which the product reaches the final consumer. An idea of how that system works in the case of the Yorkshire and Humber region is given in the boxed extract. There are variations between countries. For example, the role of the auction at the first stage of the system is strong in Scotland, but weak in Poland. In the UK the size and status of the multiple retailers have gathered strength, which has had a major influence on the way the processing sector has developed. In Mauritania the state intervenes in the marketing chain through a parastatal fish marketing organisation. In Las Palmas fish trading agencies have a fundamental part to play in buying and selling fish to the main importers in Japan and other countries. In Bangladesh the remarkable trade in fish fingerlings from the nurseries to the fish farmers throughout the country is an important part of the system. The fish businessman has to prepare a picture, or model, of the system. He does this because he can then see where his intervention is likely to be the most effective and profitable.

Market share

Figure 8.1 provides a helpful tool for analysing market share for the entrepreneur. It takes the forecast beyond mere guesswork because the quantities of fish (or whatever else is being offered to the industry) need to be estimated. It forces the businessman to think hard about realistic numbers for his own sales, when he takes his own selling price into account.

Market price

As any economist will say, price is an important determinant of market share so, if possible, the fish business needs to compare his prices with the prices offered by the competition. Another lesson of economics is that in a competitive market, if the same product is being offered, it is simply impossible to receive a higher price than the competition. It points to the need for small businessmen to offer something special, so that a better price can be expected.

Market share and other product attributes

There is also a range of other product attributes which need to be considered when examining market share. Among these are: design, packaging, presentation, appearance, accompanying use advice, availability of special requirements, delivery, hours of business, colour/flavour/odour, image, specification, payment terms, freshness, source and perhaps others. So it is also useful for the businessman to compare what he offers in respect of these attributes with the competition (Fig. 8.2).

Market trends

The fish businessman has to recognise that the market is always changing. Part of market awareness is to maintain a steady grip on the way the market is changing. So if a fish businessman already has a company up and running, or if he is thinking of starting out, the analysis of change is crucial.

An important underlying driver for change is the advance of technology. Most people would select the change in information technology as being crucial for the way that many industries have developed over recent years. This is also true of the fisheries sector. For example, fish-finding equipment has become more effective, so vessels are able to be more productive, provided fish is available. Computer-controlled systems for monitoring and managing fish farm operations are available, and can help fish farmers become more productive. Some of the advances in fish processing are noted in the boxed extract. There have also been advances in the use of computers in fish auctions.

A second major driver of change in fisheries is the regulatory system. The catching of fish is heavily regulated now and there is no realistic prospect of that burden decreasing. Fish processing is also increasingly regulated, particularly by new hygiene and safety regulations. Even fish retailing is following the same path – much to the discomfort of people who are trying to run small businesses which have little space and limited financial resources.

The third major factor comes under the heading of broad economic changes. The macro-economy is always on the move, usually implying gently rising living standards for the majority of people. This has implications, usually favourable, because of a positive relationship between income and fish consumption. At the same time tastes change and this too can impact upon the market for fish.

Investment in marketing

A hard lesson for new fish businessmen is that sales do not just happen. Investment of money and time are essential if a level of sales is going to be established.

The role of the FAO

The Food and Agriculture Organization (FAO) of the United Nations has performed an invaluable service in the fish trade through the work of its Fishery Industries Division, which is established in each of the main areas of the world. The headquarters, Globefish, can be contacted at FAO in Rome.

The role of price in marketing

Price theory

The theory of price determination is one that has interested many economic researchers. It follows that there is a considerable volume of theoretical work about how price is determined (Douglas, 1987; McAleese, 1997). The theory backs up a practical understanding of the market because it tells the fish marketing enterprise what is a feasible strategy in its own circumstances. The economic theory of markets is bounded by, on the one hand, the model of perfect competition and, on the other hand, the model of monopoly.

The assumptions underlying the model of perfect competition are: (1) many buyers and sellers; (2) a homogeneous product; (3) free entry and exit; (4) perfect information to allow rapid adjustment of price. Underlying these assumptions is the important idea of no barriers to entry. It follows that a useful check on the degree to which markets may be considered to be competitive is the extent to which there is evidence of barriers to entry. Barriers to entry arise when the *architecture* of businesses already in the industry is difficult to reproduce, when their *reputation* is deeply embedded, when *innovation* needed to get into the business is very expensive for newcomers and when existing businesses own or control *strategic assets*. Under perfect competition it is assumed that no buyer of the product is able to influence price or quantity by his own decisions. It is a well-known fact in the fish industry that this assumption is often some distance from the reality of fish markets, so, perhaps more important for this text, fish business people need to try to find ways of preventing the market approximating to perfect competition, so that they can extract a larger EAV from it. Thus, if there are barriers to entry (such as licences, skills, contacts and reputation) and they can clamber over them, then they are in business.

The assumption of homogeneity means that there is no distinction between the outputs of different producers. One of the mistakes which those without fisheries experience make when coming first to the industry is to assume that the product is homogeneous. However, one learns from bitter experience as a fish trader that the differences between products begin on the sea bed, with the variety of species, sizes and fishing methods. These variations are multiplied by the handling practices of different vessels, and then multiplied again by the handling practices of different fish merchants and the variety of presentations.

As regards price policy, if an enterprise is in a highly competitive environment and is producing a product for which there are identical or very close substitutes, it can be nothing other than a price taker. In this wide arena of the world fish market many producers are small and have no choice but to accept what the market will bear. One of the arts of marketing, however, is to find ways of differentiating the product in the minds of buyers – building architecture (or, in the marketing jargon, segmenting the market). For example, Ecuadorian shrimp exporters have successfully established a reputation for quality which lifts their prices above the average and leads, therefore to EAV.

In the case of pure monopoly the firm and the industry coincide. There are few, if any, examples of the pure monopoly on the selling side of the fisheries sector, but fish business people sometimes have to deal with monopolies and near-monopolies, such as water companies and some harbour owners. One of the intermediate cases between monopoly and perfect competition is known as monopolistic competition. It is a term reserved for those cases where although there are many firms, their outputs are not perfect substitutes. Monopolistic competition is an apt description for what goes on in many sectors of the fish industry. Fish merchants are producing similar products, and yet, through location, through personal contacts and through reputation, they are able to exert some influence on the price of their product. The description also rings true for the remaining state fish marketing enterprises which have to compete with the private sector, and yet have characteristics that enable them to offer something special to their buyers.

Oligopoly describes the case where there are only a few firms producing the product. For oligopolies, business strategy, including marketing and price strategy, is paramount. Each business has to consider very carefully what customers and competitors will do under different management decisions. Those fish marketing enterprises that face just a few buyers fall into this category.

Pricing in practice

The principle of equating marginal revenue with marginal cost (known as marginal cost pricing) remains a logical, and actually very simple, profit maximising pricing rule for businesses. In practice it works as follows. Assume that the price per kilogram of the output is $10, and the sales are 100 kg, sold in units of 10 kg. If the price is dropped by $1 to $9 per kg, it is predicted that sales will increase to 120 kg. Thus, the incremental (that is, discrete marginal) revenue from the price reduction is the new sales revenue minus the old sales revenue (120 × $9 – 100 × $10), or $80. If the cost of producing the incremental 20 kg is less than $80 per kg, then a profit maximisation policy reduces the price to gain the extra revenue. One is losing $80 or less in extra cost, but gaining another $80 or more in additional revenue.

Mark-up pricing

Mark-up pricing is also known as cost-plus pricing. The principle is that some fixed percentage is added to the average variable cost. Thus:

$$P = AVC + AVC \times X\%,$$

where P = price, AVC = average variable cost and X% = the percentage mark-up. The variable costs are those costs that would be avoided if no output were produced. The percentage mark-up can be thought of as the firm's attempt to ensure that the returns make a contribution to fixed overheads. If average variable costs were $8 per kg, a mark-up of 25% would mean that the price would be:

$$\$8 + \$8 \times 0.25 = \$10.$$

It is quite wrong to assume that mark-up pricing in practice is simply based on costs. In practice, businesses have an eye to what the market will bear when they are considering what prices to charge.

Pricing as part of marketing strategy

In the real world, fish businesses must compete nationally and internationally. It is important, therefore, for them to make use of all the marketing techniques that are available. The marketing of any commodity involves much more than manipulating the price. In economic terms marketing techniques aim at altering the structure of demand to the advantage of the enterprise in question. Price is one part of the buyer's perception of the product, but other factors, such as the location of sales, the amount and quality of advertising and promotion and the way the product is presented to the buyer, may be influential. In addition, many other factors, such as the prices of substitutes and advertising and availability of competing products, will be influential. The subject is a vast one, with both economic and marketing dimensions.

Price positioning

Price positioning is the term used to describe the choice of a price which places a product in a position in the market to reflect its bundle of attributes. A company manufacturing a range of prepared consumer fish products might be careful to price them so that the most expensive has what market research has shown to be the most appealing features.

Product-line strategy

In the case of product-line strategy, the enterprise understands that the products it produces are complements or substitutes and chooses its prices accordingly. Take the example of a fish and chip shop. Its two major products, fish and chips, are complements in the sense that a reduction in the price of one leads to an increase in its consumption and, because fish and chips are eaten together, an increase in the consumption of the other. Thus, if the business reduces the price of fish, it can expect to sell more chips, and, conversely, if it overprices chips it will reduce its fish sales. The implication for the business is that the relationship between the prices of fish and chips has to be carefully chosen.

Perhaps fish and chips also compete with other products sold by the shop. In this case the bundle, 'fish and chips', is a substitute for other offerings, such as meat pies. Thus, if the price of fish and chips rises, the consumption of fish and chips falls, and buyers then begin to substitute towards meat pies. Prices must then be selected in relation to the prices of other products so that the configuration of prices generates the best return for the businessman.

Pricing to imply quality

It is possible for some consumer goods industries (such as cars and cameras) to use price to imply some special product quality.

Commodity bundling

The pricing strategy at work in commodity bundling is the well-known phenomenon of offering one item at one price (for example, £1) and a bundle of the same commodity at less than the equivalent multiple of its price (for example, 2 for £1.50). For this to be worthwhile for the supplier the incremental benefit from selling an extra unit must exceed the incremental costs of producing it (see 'Pricing in Practice' above for a description of marginal cost pricing). Why should the consumer purchase two (or more) at a bargain price? The more the consumer purchases, the less he is willing to pay for the marginal unit – his 'reservation price' falls. However, as long as what he is actually asked to pay for more units is less than the price he would be willing to pay for the marginal units, he will buy bundles of goods. For some people the extra cost of buying a bulk package of fish will be less than the expected benefit, so they will be tempted by bulk offers, but others will prefer to buy in smaller units.

Promotional pricing

Sometimes businesses promote products by cutting the price for a period. When they do this it is in the hope that the demand is elastic and that sales volume will increase by more than the cut in price. The two factors that influence price elasticity of demand are the number of close substitutes and the proportion of consumer income spent on the good. For most (that is 'normal') goods the substitution effect and the income effect work in the same direction. Thus, a price fall induces consumers to substitute out of other goods towards the reduced priced goods and consumption will receive an extra boost from the fact that people now have more money to spend on it as a result of the price fall. So we expect promotion of goods by means of price to be most effective where goods have a number of close substitutes and/or a large proportion of consumer income is spent on them. Fish is not normally a substantial item of consumer expenditure, so the enterprises should look out for substitutes as an indicator as to whether price promotion might work. The intuition behind this is that when fish is an important item of expenditure and/or there are close substitutes, consumers will jump at the chance of a lower price.

New product pricing

Quite frequently the decision on a price for a new product will be dictated by the prices of similar products on the market. However, it is worth noting that when a product is new the scope for 'price skimming' (offering for sale at a high price) is often greater. The alternative strategy, 'price penetration', might be appropriate if a positive advantage in securing a large initial market share is identified. But the presumption is that in the case of a genuinely new product, the demand is likely to be inelastic over quite a wide range. It follows that this presents the enterprise with the means to secure EAV from the product by offering for sale at a high price.

Auctions

There are essentially two types of fish auction (see Chapter 12): price incremental and price decremental. In the former case the price is bid up by the buyers. In the latter case the price starts high and is then allowed to fall, with buyers intervening when they want to make a bid. Economic theory suggests that the latter may be more remunerative to the fishermen than the former. In the incremental case the buyer who most wants the fish only has to pay a fraction more than the second keenest buyer, whereas in the decremental case the keenest has to pay close to or at his 'reservation' price if he wants to be sure of securing the supplies he needs – he may not know what the next bidder is willing to pay so his price might be significantly above the price of the next keenest buyer. However, whether in practice there is much difference is doubtful. Information on prices in alternative markets is too widely available for fish merchants to be much influenced by the direction in which prices are moved in the auction.

Selling less popular species

Is there ever a case for businesses to sell less popular species, perhaps at a loss? First, it may be part of a bargain between the fish marketing enterprise and the fishermen. The business buys unpopular fish and in return, when supplies are tight, the fishermen supply the high-priced species for less than the going rate. Second, it may be the case that as a matter of government policy it is thought to be worthwhile to increase fishermen's incomes and fish processors are instructed or subsidised to buy. Various arguments might support this, such as the need to encourage village enterprise and development. Then the benefits accrue to the fishermen and their families, but the costs are borne by the business or the state. Third, it is possible that the business is in a position to develop new products and markets for unpopular species. For example, whilst there may be no market for shark in some Pacific Island communities, an enterprise may have the resources and knowledge to develop the market overseas.

Fourth, price policy might be used to increase fish consumption on nutritional grounds, but it is likely to be misguided. There are three main arguments against it (Timmer *et al.*, 1983; Timmer, 1987; Ellis, 1992). First, it is now widely accepted by nutritionists that poor people derive most of their basic nutritional requirements from staples. Thus, the best way of improving the nutritional status of the extreme poor is to increase the supply and lower the price of those commodities on which they most depend, not of foodstuffs that, in terms of their total food intake, are, in most cases (but not all), unimportant. Second, subsidising protein products for the poor has an opportunity cost. Given a nutritional objective, it will often make much more sense to use the resources spent on fish to boost staple food supplies rather than fish. Third, in any event, subsidising fish will frequently involve subsidising the better off who also benefit from the low-cost fish supplies as well as the poor who really need them. The welfare of the poor is likely to be better served by ensuring that fish supply is efficient and is priced at its economic cost. This is the way to make people better off.

Stockholding

Sometimes fish businesses put fish into stocks. For example, a freezing and cold-storage facility may be attached to a fish catching project with the specific objective of holding fish until the market has improved. When fish is put into cold store for later sale, the buyer ties up capital. First, this has an opportunity cost because he could be using his capital elsewhere in the economy. The second cost is the cost of freezing and cold storage. In many countries these costs are high, mainly because of electricity costs. The third cost is that of risk and uncertainty. When a decision to freeze fish is taken, the price received at a later date should reflect the loss of interest on the capital tied up, the costs of freezing and cold storage, and also the inherent uncertainty about the future movement of prices. The commercial logic of cold storage thus points to buying cheap and selling dear. It is in principle quite possible to calculate the unit costs of cold storage – the interest charge appropriate for the period of time, the freezing and unit cold-storage costs plus some allowance for risk and uncertainty. The difference between the price paid to the fish producers for the fish and the expected price to be received by the fish marketing enterprise at some future date should be equal to or more than this cost if stockholding is to be worthwhile.

Grading schemes and fish prices

There is a case for a public system of fish grading. In the case of the European Union, the Common Fisheries Policy requires the application of freshness and size grading to all the main species landed. The freshness grades (E, A and B) and the size grades (from 1 to 5 in the case of cod, but 3 or 4 for most other species) are determined by the Council of Ministers, the supreme EU authority. The economic argument for a common system of grading is that it reduces transaction costs. It is easier for consumers to know what they are buying and trade flows more easily if buyers and sellers in different countries have a common yardstick. Thus, the fish trade, lubricated by common standards, ensures that fish flows to where the demand for it is greatest, that is, where it will give the most consumer satisfaction. The case against a grading system is that the market is quite capable of estab-lishing its own standards without official intervention. Consumer tastes in regions and countries differ, but ultimately consumers' own preferences establish grades. The requirement to apply an official system adds unnecessary handling costs to fish distribution and may distort consumer preferences. A very useful role for the state is to advise fish producers and wholesalers on fish grading. This can be done whether grading practices are officially determined or whether (as in most cases throughout the world) they are based on good marketing practice. For example, if fish is to be exported to Japan from Pacific Island states, this will have implications for how it is to be handled and graded from catching through transportation to packaging. All those concerned in the distribution chain need to understand what is required of them.

Withdrawal prices and minimum prices

The EU operates a system price support for the catching sector which is based on voluntary membership of a cooperative organisation known as the fish producers' organisation. Under the system, correctly graded fish that fails to attract a buyer at some predetermined price (the 'withdrawal price') is withdrawn from sale. Subject to official certification that the fish is correctly graded and that the withdrawal is genuine, the fish producers' organisation may claim reimbursement of a proportion of the withdrawal price from official sources (see Chapters 11 and 12). The remainder of the due compensation is met by the fish producers' organisation itself from its own financial resources. The procedures for calculating withdrawal prices and the compensation payable are complex and are not strictly relevant here. The main point is that the withdrawal prices are officially determined minima which are backed up by EU financial resources.

Fish prices equalisation

Entirely self-financing schemes of price support have also been applied in Europe. These are known as *fish price equalisation schemes*. The cost of compensation for unsold fish is met by a levy on participating fishermen. The levy could be raised in a number of ways, such as a standard rate per unit of fish landed, an *ad valorem* levy, or one that differentiates between species. Fish price equalisation schemes depend for their success on the inelasticity of demand. At times of glut the fishermen's organisation, such as a fish producers' organisation, withdraws fish. The gain in total revenue as a result of the withdrawal is greater than the loss through the reduction in volume. It seems a fairly obvious way of increasing producer incomes, but this is not necessarily so. In the short-run it is true that there may be some gain, but in the long-run demand is likely to be much more elastic. In practice, if fish merchants repeatedly find themselves deprived of low-cost fish they will sell less than they would do with more fish available. To some extent supply creates its own demand which is encouraged by plentiful supplies of low cost fish and continually depriving the market of this supply may inhibit market development.

Price stabilisation

Price stabilisation involves buying fish up at times of glut and selling it at times of scarcity to smooth out price fluctuations, and now takes place in the case of some species with EU support. There are a variety of complex reasons why such schemes are not generally successful. Two may be mentioned. It is very difficult indeed to persuade fishermen to forego income when prices are high so that there are resources available for purchasing fish at times of glut. This is not surprising. In general, people prefer to make their own decisions about how they allocate their incomes, so if the fishermen want to they can save surpluses earned at times of high prices to provide a cash reserve for later on. A preferred solution, then, may be to ensure that savings institutions are available for fishermen so that they can save and earn interest as a safeguard rather than a price stabilisation scheme. Second, the costs of establishing and running price stabilisation may be greater than the benefits (if any) it can offer.

Market Information Systems

Market information systems (MIS) have four components.

The system of management accounts

Certain information on costs and sales will be collected for routine audit purposes. The accounting system for audit purposes will have been designed by accountants to ensure that the transactions of the organisation have been properly conducted and to permit the calculation of profit and loss, flow of funds and a balance sheet. In many organisations the system will be computerised.

It is most unlikely that the accounting system for audit purposes will yield very much management accounting and sales information of the kind required, although it will almost certainly be possible to attach an additional information requirement to the accounting system to make it produce useful data. For example, the auditors will want to see sales invoices for wholesale sales of fish and matching transactions in the books and in the cash balances, but unless the system has been designed to generate marketing information the sales invoices will not necessarily include data on prices of each species, sizes and quality.

In respect of retail sales, a similar point may be made. Auditors look for a consistent paper trail for the money received for the sale of fish, but they are not interested in who buys fish, how much they buy, their income and so on. The marketing information will have to be added to the accounting system.

Market intelligence

The second component of the MIS is market intelligence. This refers to the collection and use of information about the wider market environment. For example, the FAO publication *Infofish Trade News* and the regular information sent out by the Forum Fisheries Agency and other similar industry data would be included under this heading. What a fish marketing enterprise chooses to collect obviously depends on its own situation. Most fish marketing enterprises seem to be reasonably well aware of what information can be gleaned by subscription to reports and journals.

Market research

The third component of the MIS is market research. This is the result of careful work with a much more specific focus than market intelligence. The commercial, profit-maximising rule for undertaking market research is that the incremental expenditure on the research should continue to the point where it is equal to the incremental benefits of undertaking the research. In many fisheries situations far too little research into markets is undertaken prior to decisions, with the result that bad decisions are made.

  For example, assumptions are made that more fish can be sold at a given price based on very cursory observations of a market. More thorough market research would have shown the segmentation and limits of the market, and provided business planners with a much more realistic picture of the possibilities for additional fish sales. Another mistake that is often made is that it is assumed that the methods of market research are inappropriate for developing countries. In fact, given that the costs of mistakes are often more serious in developing countries than in developed, it will often be worthwhile undertaking the necessary expenditure to get the marketing picture as accurate as possible. Some questions are to see how people spend their incomes, the importance of expenditure on fish, the scope for substitution between fish and other commodities and attitudes towards fish among different social groups.

Market modelling

The fourth component of the MIS is the development of techniques for modelling markets. Models of markets are increasingly easy to use with the widespread use of micro-computers. There are a number of statistical packages on the market which will help fish marketing enterprises to make more use of such data that they collect under the first three components.

The above headings for an MIS are a starting point for thinking about how market information should be collected and analysed. In a small organisation it may be just part of one person's job to undertake the supervision of the MIS, but in a larger organisation more personnel will be required. Recalling the point that incremental expenditure on gathering and analysing information should be undertaken up to the point where it is equal to the incremental benefits resulting from it, the scale and organisation of the MIS is a strategic management decision (Crimp, 1990; Kotler, 1997).

Summary

Market analysis

Identifying a market segment

The term marketers use for identifying a group of potential buyers is 'market segmentation'. There are many different segments in a marketplace and it is important to give hard thought to what they might be.

Identifying customer needs

Once a suitable segment of the market is identified the next step is to find out what this group of people or businesses really needs from the fish product. Market size, structure, the influence of price and other factors on market share need to be considered.

Market trends

The fish businessman has to recognise that the market is always changing and these changes need to be identified. Important factors include: (1) technological change, which alters fish catching and fish farming as well as fish processing and transport in dramatic ways; (2) the regulation of fisheries, much of it emanating from the EU; (3) macroeconomic changes – the economy is always on the move, in developed countries usually implying gently rising living standards for most people.

Investment in marketing

A hard lesson for new fish businessmen is that sales do not just happen. Investment of money and time are essential if a level of sales is going to be established.

The role of the FAO

The FAO has made a very useful contribution to the international fish trade by supplying up-to-date information, and therefore enhancing the transparency of markets. It does this through its Infofish network. The headquarters of the network is known as Globefish and is based at FAO in Rome.

Price theory

The assumptions underlying the model of perfect competition are as follows: (1)

many buyers and sellers; (2) homogeneous product; (3) free entry and exit; (4) perfect information to allow instantaneous adjustment of price. In perfect competition it is assumed that no buyer of the product is able to influence price or quantity by his own decisions. Fish business people should try to find ways of preventing the market approximating to perfect competition, so that they can extract a larger EAV from it.

Pricing in practice

The principle of equating marginal revenue with marginal cost is a logical profit-maximising pricing rule for businesses.

Mark-up pricing

Mark-up pricing is also known as cost-plus pricing. The principle is that some fixed percentage is added to the average variable cost.

Pricing as part of marketing strategy

Price is one part of the buyer's perception of the product, but other factors, such as the location of sales, the amount and quality of advertising and promotion and the way the product is presented to the buyer, may be influential. A selection of some of the ways in which price is used to support marketing strategies follows:

- Price positioning
- Product-line strategy
- Pricing to imply quality
- Commodity bundling
- Promotional pricing
- New product pricing

Auctions

There are essentially two types of fish auction: price incremental and price decremental.

Selling less popular species

The disposal of less popular species at a price that the fisherman or fish farmer regards as reasonable may be part of an implicit or explicit bargain between the fish marketing enterprise and the fisherman. The business buys unpopular fish and, in return, when supplies are tight, the fishermen supply the high-priced species for less than the going rate. Second, it may be the case that as a matter of government policy it is thought to be worthwhile to increase fishermen's incomes and fish processors are instructed or subsidised to buy. Third, it is possible that the business is in a position to develop new products and markets for unpopular species. Fourth, price policy might be used to increase fish consumption on nutritional grounds.

Stockholding

Sometimes fish businesses put fish into stocks. The commercial logic of cold storage points to buying cheap and selling dear.

Grading schemes and fish prices

The economic argument for a common system of grading is that it reduces transaction costs. It is easier for consumers to know what they are buying and trade flows more easily if buyers and sellers in different countries have a common yardstick.

Withdrawal prices and minimum prices

The European Union operates a system of price support for the catching sector which is based on voluntary membership of a cooperative organisation known as the fish producers' organisation.

Fish prices equalisation

The cost of compensation for unsold fish is met by a levy on participating fishermen. The levy could be raised in a number of ways, such as a standard rate per unit of fish landed, an *ad valorem* levy, or one that differentiates between species.

Price stabilisation

Price stabilisation involves buying fish up at times of glut and selling it at times of scarcity to smooth out price fluctuations, and also forms part of the EU Common Fisheries Policy.

Market information systems

There are four components to Market Information Systems: (1) The system of management accounts; (2) Market intelligence; (3) Market research; and (4) Market modelling.

APPENDIX: THE HACCP SYSTEM

EU Council Directive 93/43/EEC of 14 June 1993 on foodstuffs hygiene urges all food businesses to develop a Hazard Analysis Critical Control Point (HACCP) system. The HACCP concept first came into use in 1959, when it was a means of ensuring that food for astronauts was safe to consume. It was first applied to microbiological hazards only, but has since been developed and extended to include physical and chemical hazards. In the future consumers can reasonably be expected to be strongly disposed towards food products that combine the classical hygiene requirements with a HACCP system.

The seven principles of HACCP

(1)   The identification and analysis of hazards in the vertical chain.
       This principle applies to the detection of hazards in the area of application, or the sphere of influence of the business. If an operator can exert an influence over raw materials then he should do so along the vertical chain. If he buys in his ingredients the purchaser should ensure that hazard analysis has been carried out. A systematic assessment of the ingredients and constituents should be carried out. This analysis should include physical and chemical hazards as well as biological. The hazard analysis is based on product descriptions containing information on microbiological, chemical and physical parameters and variables. Flowcharts of the entire process are used for hazard analysis.

(2)   The identification of critical control points which must be monitored to avoid the occurrence of hazards.
       A control point becomes critical if contamination might increase to an unacceptable level and which might not be eliminated later in the production process (such as by high-temperature cooking).

(3)   Laying down values for critical limits which must be observed to control the hazards at each critical control point.

(4)   The introduction of a surveillance system for the regular monitoring or observation of the critical control points.
       The monitoring system constitutes a plan under which the Critical Control Points (CCPs) and corresponding limit and/or guide values are examined and/or observed. The monitoring is carried out using microbiological, analytical and/or sensory methods. Microbiological tests are rarely suitable for monitoring a CCP as they take a long time to complete. Analytical and sensory procedures are preferable and the values measured with them can be used as indicators, yielding information on the microbiological status. Continuous monitoring is always preferable to random sample testing but is

not always possible. Where random sample testing has to be used the test parameters and conditions must be planned in such a way that the values measured are statistically significant. In principle, however, even in a Statistical Process Control scheme, none of the values measured should be above the critical limit values. It is essential to use suitable measuring instruments which ensure accuracy of the results. Adequate documentation is also an essential component of the monitoring procedure.

(5) Laying down corrective measures which should be taken whenever an inadmissible deviation is recorded at a critical control point.
The measures in question should limit any hazard which has arisen and should ensure safe management (labelling and use) of the products. Corrective measures should be drawn up for each CCP as the methods of labelling and use have to be specifically adapted for every hazard. The corrective measures must also deal with the question of how the process can be brought under control again. The appropriate measures for this purpose should be documented.

(6) Laying down procedures for verification of the correct operation of the HACCP system (functional control system).

(7) Setting up a system for the effective management of the documentation relating to the HACCP plan (data collection/organisation of documentation). The records should contain information on: ingredients, intermediate and final products, processing steps and process parameters, packaging, storage and distribution, corrective measures, deviations in the process or product, and verification. The documentation has to be available for the appropriate authorities.

(Adapted from Lebensmittelqualitat Sachsen LQS, 1996.)

Chapter 9
Exporting Fish and Fish Products

Exporting is often good for fish businesses. It usually results in the producers earning more money than they do from selling into the domestic market. It therefore adds value to the business. Moreover, exporting forces the exporter to pay special attention to quality and keeping costs down. It invariably involves overseas visits so nationals are exposed directly to new ideas and the requirements of changing markets. So why do businesses make the decision to start exporting fish or fish products?

Sometimes the pressure to export arises because of surplus production. In the jargon of economics the export market is seen as a 'vent for surplus'. Many countries are relatively small and have markets that are limited by size and income. So after the implementation of a successful expansion programme the volume of output exceeds the quantities that can be sold at a viable market price. The company has the option of scaling back or finding new markets abroad. The company may also decide to switch production from domestic markets because it is profitable to do so. Although the domestic market needs have not been fully met, exports are seen to be more lucrative. Indeed, quite frequently the price of good quality fresh fish on the main metropolitan markets can be several times the domestic price. It is worth recalling that fortunes have been made by entrepreneurs who have been able to spot the missing link between supplier and a market.

The fish product which the potential exporter wishes to export may not be very well known, but he should not be discouraged on this account. If it is not well known, but palatable, it may be difficult to establish in advance of a trial whether there is a market for it. In general, the author of this text takes a cautious view of market potential, based on an understanding of the economics of consumer behaviour, which, in matters of food, displays remarkable stability. However, sometimes businessmen are surprised by how well a new product can take off. The history of the fish industry is littered with examples of fish products which people thought were not acceptable to the consumer but which became popular. Two European examples – scampi (nephrops) and smoked mackerel – and one South Pacific example – deep-water snappers – give support to the case. So the lesson is that potential exporters should not be discouraged. They may fail, but they will not know they have failed until they have tried.

The complexity of the world fish market is noted in Chapter 8. The following list of traded categories of shrimp demonstrates to the uninitiated just how complicated the market can be:

- Whole, shell-on, raw, frozen
- Whole, shell-on, cooked, not frozen
- Whole, shell-on, cooked, frozen
- Headless, shell-on, raw, frozen
- Headless, cooked, peeled, frozen
- Headless, peeled, undeveined, raw, frozen (PUD)
- Headless, peeled, deveined, raw, frozen (P&D)
- Headless, cooked, peeled, canned.

Exporting as investment

Setting up a programme of exports should be conceived of as an investment which yields a return over a number of years. It cannot really be done unless the exporter is prepared to spend some money investigating markets, obtaining appropriate packaging, making contacts, buying publicity and setting up a system which will transport and sell the product. It is also a high-risk venture so conventional investment appraisal criteria indicate that the return should be above average to compensate for the high risk. High risk is a consequence of the intangible nature of much of the investment and the strong possibility of failure, so an exporter should expect to earn good profits within a year or two of the initial investment. Thinking of exporting as an investment helps to show that (1) the exporter must be prepared to spend if he is to win a return and (2) success will not necessarily be instant.

This chapter does not go into detail of the kinds of fish and shellfish most suitable for the export market. It hardly touches on technical aspects, such as temperature, the use of ice and the transportation of live fish and shellfish. It is intended as an introduction to a potential new line of business for people selling on the domestic market.

A step-by-step guide to exporting is given in Fig. 9.1. Furthermore, readers may find the Appendix useful to identify terms and abbreviations used in the exporting trade.

Quality of exports

Chapters 2 to 5 discuss the formulation of strategy in fish businesses. They emphasise the importance of establishing a relationship with the buyer (architecture) and building up a good reputation. Seller and buyer are often involved in what might be analysed as a repeated game (see Chapter 5) within which returns to both may be improved by cooperative strategies. If one party to the game defects the relationship rapidly degenerates into a Prisoner's Dilemma, and both are the losers.

	Visit export market
	Make informal contact with buyer
	Gain written agreement with buyer
	Obtain export licence if necessary
	Obtain health certificate if necessary
	Book shipping or aircraft space
	Check with bank that LC is through
	Check terms of LC
	Prepare invoices – ensure enough originals
	Prepare certificate of origin
	Pack fish
	Notify customs
	Prepare and lodge bills of lading or airway bills
	Insure if necessary
	Forward documents to buyer direct or through bank
	Send goods to ship or aircraft

Fig. 9.1   Exports step by step.

At whatever end of the food business, a reputation for quality is essential. One or two examples of substandard products, particularly if they result in illness for some consumers, can have very serious and long-term effects on a developing market. This points to a role for governments in the regulation of exports. Some options for the institutional setting are outlined below.

The role of the state in promotion of exports

State marketing authority

The government decides to set up an institution (company, statutory body, or an arm of the civil service) which has a monopoly of fish exports. Control is absolute and quality is assured provided the institution abides by certain standards. Economists have been known to approve of this approach because the power to control the buying price from the catching sector resides with the institution. It can, therefore, adjust its price up or down to discourage or encourage fishing and therefore to regulate fishing effort. This approach is also a means by which a state which is short of tax revenue can divert some of the surplus from fisheries resources into the state coffers. It buys cheap and sells dear on international markets. For many years, Mauritania operated the 'Société Mauritanienne de Commercialisation de Poisson' (SMCP) precisely to ensure that the rent from fishery did not all disappear into the hands of the private sector. However, the SMCP did not have fisheries management as part of its remit.

The state may also decide that monopoly control is unnecessary, but that state investment is required to develop exports. Vanuatu's 'Port Vila Fisheries Limited' is an example of such an entity. The company was set up in 1983 by the state to market fish both at home and overseas. Others were also at liberty to do so, but the state-owned company dominated domestic and export fish marketing. In the end, the state did come to the view that these essentially commercial activities are better left in the hands of the private sector, so it was finally sold to the private sector in 1996.

Fish producers' organisations

The concept of the FPO is discussed in Chapter 12. FPOs are established under state protection and support. Some are becoming increasingly involved in commercial fish marketing, including exporting.

Inspection service

The state leaves exports to the private sector but appoints trained inspectors to ensure that quality standards are maintained. Concern about food safety leading to the development of statutory hygiene standards will result in a significant expansion in this approach.

Training

In the past, governments may have considered that training the staff from exporting companies might be sufficient to ensure that the industry is self-regulating. The UK might in the past have taken that view, but increasingly statutory regulation has been imposed. This has actually reinforced the need for training in order that staff know their way around the regulations and how they should be implemented.

Market research

It is too easy to assume that investigating foreign markets will be too expensive for the small businessman. It need not necessarily be so, as is recognised, for example by the Midland Bank (1988, p. 7). If the export volume is likely to be low relative to the size of the market it is hoped to sell into, it might be possible to spend a great deal less than the advisory books, erring on the side of caution, suggest. It is possible to start exporting into an established market without too much cost. In the fish industry exporters actually have something of an advantage because, in general, fish is in short supply and buyers are often very happy to see newcomers coming in with a new source of supply. Exporting then becomes a problem of establishing the right contacts and making sure the paperwork is correct.

Numerous different techniques are now available to market researchers, such as sample surveys, focus groups and taste panels (see Chapter 8; Chisnall, 1986 and 1991; Crimp, 1990). It is important for any business thinking about entering a new market to be aware of these so that an informed judgement on market assessment can be made. However, it may also be possible to assess markets through personal contacts. A prospective exporter should begin his attack by studying official sources of information. (A number of these are available in the UK and are listed in Chapter 7.) Embassies, government departments, statutory bodies and international organisations often have data which can be accessed at low cost, or sometimes no cost. It will often be possible to put together quite a detailed picture of the market for fish in a country from such sources.

Official sources may provide a list of persons or companies who are willing to act as agents for the exporter. However, a potential pitfall is that they do not know who is likely to be trustworthy. They may not necessarily have detailed knowledge of the various import regulations of their country. The trade department may have good knowledge of trade rules and say that the product would be accepted, but is not fully conversant with hygiene rules. The author is aware of two cases where exporters were encouraged to go ahead with consignments, having assumed official sanction for the trade on the basis of official contacts, only to run foul of hygiene regulations when the fish reached the recipient country. This can be a very costly error for the exporter.

The next group of organisations which can be helpful are trade associations. Again a word of caution is in order. Trade associations are often very small with limited resources. They aim to represent the interests of the members so they will not give advice which undermines the interests of any of them. Also, they may find it difficult to give advice which improves the competitive position of one of their members compared to another. They are also unlikely to divulge information on which of their members are the most trustworthy.

The potential exporter must also be prepared to go out and talk to people. He can go through the yellow pages of the telephone directory to find contacts. He can visit the retail markets and pick up leads from fishmongers. He can knock at the door of the large fish processing companies. He can visit port and inland wholesalers. After a few days he will develop an understanding of the market. He will also be able to make a judgement on the price he can expect to receive for his fish from wholesalers. As a very rough and ready rule-of-thumb this is likely to be roughly half the advertised retail price for similar kinds of fish in the shops.

The internet is also a useful source of information about fish markets. Buyers and sellers post their requirements and offers on various web sites. The general pattern is that once established between traders who know each other internet and e-mail trading is common. However, in the first stages of trade there is really no substitute for the personal visit.

Figures 9.2 and 9.3 offer some suggestions of who to see and what to ask for.

	Buyer's embassy or high commission
	Ministry responsible for fisheries in buyer's country
	Ministry responsible for trade in buyer's country
	Ministry or department responsible for food hygiene
	Customs and excise authorities
	Relevant statutory bodies
	Trade associations
	Direct buyers at their premises
	Import/export agents
	Merchants at wholesale markets

Fig. 9.2   Checklist of who to see.

The agent

At the early stage of fish exporting it may turn out that the simplest arrangement is to appoint an agent in the recipient country. Well-known exporting businesses often receive unsolicited offers from individuals or companies to act as agents.

The main channels by which suitable candidates can be found are:

- Approach by prospective agent to exporter
- Recommendation through friendly contact
- Chambers of Commerce
- Official (embassy) commercial representatives, e.g. commercial attaché
- Trade associations
- Advertising by potential agents in trade journals
- Advertising by the exporter seeking an agent.

The following is a checklist of qualifications which the first-time fish exporter should require of his agent:

	Names, addresses, fax numbers and e-mails of Direct buyers
	Importing agents
	Statutory bodies and contacts
	Ministry of Trade and contact
	Ministry of Health and contact
	Ministry for Fisheries and contact
	Authority concerned with food hygiene and contact
	Customs authority
	Trade associations and contact
	Import duties
	Health regulations (in writing)
	Packaging requirements (legal and market)
	Prices
	Presentation requirements
	Fish transport
	Copies of relevant reports
	Export procedures

Fig. 9.3   Checklist of what to ask for.

- Knowledge of local (market) language
- Knowledge of exporters' language
- No agency in competition with exporter
- Complementary agencies (e.g. fruit and vegetables) may be desirable
- Enthusiasm for fish
- Knowledge of and contacts in the local trading community
- Scope for continuity of the person within the agency responsible for the exported fish.

When the exporter has found a suitable agent he should enter into an agreement with him covering the following terms:

- The exporter's name, address, fax number and e-mail
- The agent's name address, fax number and e-mail
- The territory to which the agency relates
- The products covered by the agreement
- The date of commencement and, if relevant, the date of termination of the agreement
- Other arrangements for terminating the agreement
- The rate and method of assessment of commission and other remuneration
- Special details, if any
- The legal background to the agreement and provisions for arbitration.

Generally, an exporter will be advised to employ a lawyer to draw up the agreement between himself and the agent. This is obviously desirable as it gives both parties more security. However, it is quite possible to reach morally and legally binding agreements without the use of lawyers if there is a good trusting relationship between the exporter and his agent. It may not always be necessary to go to the expense of obtaining legal advice.

Samples

Agents and potential buyers are the best source of information about the form and presentation of samples. An exporter is expected to provide samples and maintain the same presentation and quality in his shipments. Some buyers are prepared to pay for samples. However, it is not necessary to bring samples on the first exploratory trip to a country. It is perhaps more impressive to be able to send for samples of products, especially if they are of fresh fish, so that they arrive by air within a day or two.

Printed literature

Printed literature serves two purposes. It gives the potential buyer an opportunity to find out about the fish scene in the exporter's country. It tells him how and where the fish is caught, who caught it and where it is processed and packed. It tells the buyer what kind of country is sending the fish and if future supplies are likely to be reliable. A well-put-together brochure is a useful selling aid.

The brochure also tells the buyer what kind of people he is dealing with. If it is an attractive, professional document, with no spelling mistakes or printing errors, written in the language of the buyer, with good quality photographs, it is an excellent advertisement for the exporting country and its fish products.

The brochure should contain as a minimum:

- Photographs showing the fish in the form it is to be sold
- Photographs summarising the catching and processing

- Photographs showing fish handling conditions
- Text summarising the economy plus the role of the fish industry within it.

Packaging

The exporter needs to take advice on and pay very close attention to packaging. The packaging of traded fish is changing all the time and whatever is used needs to meet the buyer's requirements.

Labelling

The buyer's country will probably have legal labelling requirements. The exporter must make careful note of these so that when the fish arrives in the buyer's country it is not held up for technical reasons. The label will probably include the scientific and common names of the species and the weight. The shipper or airline will also have their requirements to be met. These should be established before exporting.

Price list

A price list should give:

- The period of validity of the quotation
- The units in which the prices are quoted (e.g. $/kg)
- The packaging details (e.g. packed, layered fish in 10, 20, 25 or 30-kg cartons)
- Details of FOB (free on board), CIF (cost, insurance and freight), C&F (cost and freight), etc
- Terms of payment
- Scientific names of species
- Common names of species
- Presentation (e.g. whole, gilled and gutted, fillets, steaks, etc.).

Documents and endorsements which may be needed

Export licence from exporter's country

In some countries there are controls on exports and an export licence is required. There may be bureaucratic or tax reasons for this. But actually it may also be useful to the exporter because it provides an official statement that what is being exported is what the seller says it is.

Import licence

The buyer may need authority to import the product.

Commercial invoice

A commercial invoice is a claim by the exporter for payment for the goods. It should include a detailed description of the goods, the unit prices, the total weight and terms of payments, as well as packing and shipping details. Several copies of the invoice should be made and the top three should be identified as originals so that the documentary requirements of the bank are met.

Customs invoice

A customs invoice is not required by all countries. Some countries insist on it so that the authorities and the overseas buyer have a further check that the correct duty has been levied.

Certified invoice

Not all countries require a certified invoice, a commercial invoice bearing a detailed statement of value and origin, signed by the exporter. Certification is the process by which the authorities of the buying country, probably its consulate, legalise the commercial invoice before it is sent.

Packing list

A packing list may be required by the shipper to show the number and weight of cartons. One copy accompanies the goods and another accompanies the commercial invoice.

Bill of lading

A bill of lading is the receipt for goods shipped; it is a legal document and only applies to goods shipped by sea. It is a document of title which means that the shipping company named on the bill of lading has the legal right to possess the goods. It is usually made out in a set of three originals and several copies.

Combined transport document

If part of the journey is by some means other than sea, a combined transport document verifies the responsibility of the carrier for the goods.

Through container document

A through container document is similar to the combined transport document, except that under this the liability of the carrier is only for the land journey between ports.

Railway consignment note

A railway consignment note tells the exporter under what terms his goods are travelling by rail. It is not a document of title and is not negotiable, unlike bills of lading and similar legal documents.

Airway bill

An airway bill does not correspond exactly with a bill of lading because it is non-negotiable. It is the documentary evidence that the airline accepted the goods and transported them. Unlike the bill of lading, a legal transfer of the goods from one carrier to another would not be possible.

Certificate of origin

A certificate of origin is often required by the importing country, particularly if some exemption from duty is being claimed.

Certificate of health, quality or inspection

The authorities in the importing country may require a certificate of health, quality or inspection as verification that the product meets certain standards (for example, see Brooker, 1981).

Insurance certificate and/or policy

Whether an insurance certificate or an insurance policy is required depends on the contract. Usually the banks only require a copy of the certificate, but a copy of the policy may be required, so it is very important for the exporter to check this.

Customs declaration

The exporter may be required to complete a customs declaration before the goods leave the country and, if applicable, to pay the duty. The importer will also be required to complete a customs declaration for his own customs.

Export finance

Whatever is exported there are four main methods of payment.

Payment in advance

Payment in advance is known as cash with order (CWO). If the buyer can be persuaded to part with his money at the time of the order this obviously provides the best possible security. However, it is very rare as it implies that the buyer is extending credit to the exporter.

Open account

An open account offers the least security to the exporter. It simply means that the exporter sends an invoice to the buyer. The buyer undertakes to send the money within an agreed period. Although banks will discourage this method of operating it is quite common, especially when the sale is to a government.

Bill of exchange

An exporter can send a bill of exchange for the value of the invoice for goods for export. He sends it through the banking system for payment by the buyer when it is presented. The exporter prepares the bill of exchange (which looks something like a cheque). The buyer receives it through the banking system and writes an acceptance across it and thus agrees to pay by the due date. The bill is called a sight draft if it is made out to be payable at sight or on demand. If it is payable in the future it is called a term draft. The period of credit identified by the bill is known as the tenor of the bill. Thus, the exporter has good control over his exports. Until the bill has been *accepted* (i.e. paid) by the buyer the goods cannot be released. The goods may be delivered and held in a bonded warehouse, but they are not released to the buyer until he has accepted the bill. A period of credit (i.e. a delay in payment) can be built into the terms of the bill.

The normal procedure for payment of bills of exchange is known as documentary collection. The exporter sends the bill to the buyer through the banking system (i.e. from the bank in his own country to a correspondent bank, known as the collecting bank, in the buyer's country). The shipping documents are sent with the bill so until the bill is accepted by the buyer he cannot get hold of the relevant shipping documents, such as an original bill of lading, so he cannot claim title to the goods. When he has accepted the bill, by signing his name across it, he has committed himself to paying and he can then claim the goods. If no documents are required the procedure is known as clean bill collection.

Letters of credit

The procedure for the issue of and payment under a letter of credit (LC) is as follows:

(1)   The exporter and the foreign buyer conclude a sales contract. This would take place in person, over the telephone, fax or e-mail. Among other things, it would specify payment arrangements, such as by LC. Normally an LC is irrevocable. This implies that when the buyer's conditions have been agreed by the exporter they constitute a definite undertaking by the buyer's bank to pay, provided the buyer's conditions are met.

(2)   The buyer (the importer) instructs his bank to provide credit in favour of the exporter. The credit is essentially a promise to pay provided the agreed conditions are met.

(3)   The buyer's bank sends the letter of credit to the correspondent bank in the exporter's country.

(4)   The correspondent bank in the exporter's country tells the exporter that the LC has arrived. The exporter can then read the LC and check that the details in it conform exactly to the export consignment and documents.

(5)   All being well, the exporter sends off the fish.

(6)   He then proves to the correspondent bank that he has done so by presenting the shipping and other required documents.

(7)   The correspondent bank then pays the exporter, or the LC may allow for a delay in payment.

(8)   The correspondent bank sends the documents to the bank in the buyer's country.

(9)   This bank then checks the documents and sends the money to the correspondent bank.

(10) The buyer's bank then debits the buyer.

(11) The buyer's bank releases the documents to the buyer, enabling him to claim the goods.

(12) The buyer obtains the goods.

(13) In the case of fish, rapid transit will be essential so very often the documents will have already been sent to the buyer or will have been sent with the goods so that he can pick up the fish without any delay.

Common errors on letters of credit

When the exporter's bank has the LC it contacts the exporter. The exporter then presents the documents needed to prove that the fish has been sent. Very often the documents do not coincide exactly with what is stated on the LC. Common errors are:

* The LC has expired. In other words, the consignment is sent after the expiry of the LC.

- The shipment is late arriving in the buyer's country.
- The documents are presented after the date stipulated in the LC.
- The LC specifies the local or common name for the fish in the buyer's country. The exporter has a different name on his documents.
- The LC specifies that original documents must be presented, but the exporter only has copies.
- The LC includes problematic clauses which cannot be met by the exporter. For example, it might say 'trans-shipment prohibited' when this is impossible, or it might say 'part consignment not allowed' when the exporter is compelled to split the consignment.
- The bill of lading is claused. This means that the shipper thinks that the consignment is inadequately packed or is uncertain whether the consignment has been loaded. A clause is then added to the shipping document to this effect. The bank may refuse to pay if the LC stipulates 'clean' bills of lading but they are 'dirty' or 'unclean', i.e. claused in the way described.
- The goods are sent between ports other than those stated on the LC.
- The wrong insurance documents (e.g. the policy rather than the certificate, or vice versa) are presented.
- Insurance arrangements do not coincide with those on the LC.
- Spelling mistakes.
- Weight discrepancies.
- Money discrepancies.
- Short shipments.
- Missing documents.
- Other inconsistencies.

In summary, the LC must coincide exactly with the terms of the contract between the exporter and buyer and all the documents stipulated in the LC must be presented in the correct form. Figure 9.4 is a useful checklist to ensure that the LC is correct.

Special features of export to the EU

Marketing fish in the EU

Fish in the EU consists of:

- Fish and fishery products
- Live bivalve molluscs, echinoderms, tunicates and marine gastropods
- Aquaculture animals and products, including eggs, gametes and other products for human consumption
- Snails and frogs' legs
- Fishmeal.

	Correct spelling
	LC describes exactly the agreed contract
	Invoices, packing list, airway bills or bills of lading and insurance documents coincide exactly with LC contents
	LC is irrevocable
	Money due is paid in agreed manner
	Time delay in payment acceptable
	Place of payment acceptable
	Dates acceptable

Fig. 9.4 Letters of credit (LC) checklist.

International trade is included in the Common Fisheries Policy's marketing components. It has four main instruments:

(1)   The application of common standards for marketing products.
(2)   The recognition of Fish Producers' Organisations (FPOs). There are now about 150 officially recognised FPOs within the EU.
(3)   The price regime, consisting of a withdrawal price system and some other compensatory mechanisms.
(4)   Trading arrangements with third countries – to ensure that the market requirements of the EU are met without damage to EU producer interests.

Rules governing fish and fish products

Exports to the EU have to meet the general requirements of the Common Fisheries Policy. Specifically, the appropriate tariff is payable. The products have to follow the authorised procedures and health, labelling and packaging requirements need to be met. These are set out in detail in *Guidelines for Fish Exporters, Requirements for the European Union Market*, Eastfish Fishery Industry Number 20.

The EU has defined a list of approved third countries from which imports are accepted. Imports from these countries must:

● be accompanied by a health certificate;
● be from a list of approved establishments or factory vessels – such licensing to be carried out and monitored by the recognised authority in the country concerned;
● carry an identification mark with the licence number of the source of the product so that it can be traced.

For other countries, the businesses themselves must be approved and fish must carry similar certification.

Countries with authorised establishments

Authorised establishments from Bangladesh, Belize, China, Costa Rica, Croatia, Cuba, the Falkland Islands, Fiji, Greenland, Guatamala, Honduras, India, the Maldives, Mexico, Namibia, Panama, Poland, the Seychelles, Slovenia, Suriname, Switzerland, Togo, Tunisia, Turkey, the United States, Venezuela and Vietnam may export to the EU. In these cases the authorities have licensed specific companies as meeting hygiene standards which at least match those of the EU and supply them with the necessary certificate.

Single authorised establishments/factory vessels

Where the competent authority is not able to supply guarantees for the whole country, single entities can be approved for export to the EU.

Summary

Exporting is good for businesses. It often results in the producers earning more money than they do from selling into the domestic market. It therefore adds value to an economy. Moreover, exporting forces the exporter to pay special attention to quality and keeping costs down. It invariably involves overseas visits so nationals are exposed directly to new ideas and the requirements of changing markets.

Exporting as investment

Setting up a programme of exports should be conceived of as an investment which yields a return over a number of years. It cannot really be done unless the exporter is prepared to spend some money investigating markets, obtaining appropriate packaging, making contacts, buying publicity and setting up a system which will transport and sell the product.

Quality of exports

At whatever end of the food business, a reputation for quality is essential. One or two examples of substandard products, particularly if they result in illness for some consumers, can have very serious and long-term effects on a developing market.

Market research

Most of the published advice on market research and exporting encourages businesses to spend. It is possible to start exporting into an established market without too much cost. In the fish industry exporters actually have something of an advantage because, in general, fish is in short supply and buyers are often very happy to see newcomers coming in with a new source of supply.

A prospective exporter should begin his research by studying official sources of information. The next group of organisations which can be helpful are trade associations. The potential exporter must also be prepared to talk to people in the industry. In addition, the internet is a useful source of information about fish markets.

The agent

At the early stage of fish exporting it may turn out that the simplest arrangement is to appoint an agent in the recipient country. Well-known exporting businesses often receive unsolicited offers from individuals or companies to act as agents.

Samples

Agents and potential buyers are the best source of information about the form and presentation of samples.

Printed literature

Printed literature serves two purposes. It gives the potential buyer an opportunity to find out about the fish scene in the exporter's country. It tells him how and where the fish is caught, who caught it and where it is processed and packed. It tells the buyer what kind of country is sending the fish and if future supplies are likely to be reliable. A well-put-together brochure is a very useful selling aid.

Packaging

The exporter needs to take advice on and pay very close attention to packaging. The packaging of traded fish is changing all the time and whatever is used needs to meet the buyer's requirements.

Labelling

The buyer's country will probably have legal labelling requirements.

Price list

A price list should give:

- The period of validity of the quotation
- The units in which the prices are quoted (e.g. $/kilo)
- The packaging details (e.g. packed, layered fish in 10, 20, 25 or 30-kg cartons)
- FOB, CIF, C&F etc
- Terms of payment
- Scientific names of species
- Common names of species
- Presentation (e.g. whole, gilled and gutted, fillets, steaks, etc.).

Documents and endorsements which may be needed

- Export licence from exporter's country
- Import licence
- Commercial invoice
- Customs invoice
- Certified invoice
- Packing list
- Bill of lading
- Combined transport document
- Through container document
- Railway consignment note
- Airway bill
- Certificate of origin
- Certificate of health, quality or inspection
- Insurance certificate and/or policy
- Customs declaration

Export finance

There are four main methods of payment: payment in advance, open account, bill of exchange and letters of credit.

Special features of export to the EU

International trade is included in the Common Fisheries Policy's marketing components. Exports to the EU have to meet the general requirements of the Common Fisheries Policy. Specifically, the appropriate tariff is payable. The products have to follow the authorised procedures, and health, labelling and packaging requirements need to be met.

APPENDIX: GLOSSARY OF TERMS AND ABBREVIATIONS USED IN EXPORTING

Aar	Against all risks (insurance term)
A/c	Account
A/D	After date, alternate days
ad val	*ad valorem* (an ad val duty is levied on the value of goods)
AR	all risks
AWB	airway bill
B/D	bank draft (a cheque drawn on a bank)
b d i	both days inclusive
B/E	bill of exchange (a promise to pay sent by the seller through the banking system for the buyer to accept)
B/L	bill of lading
CAD	cash against documents
CBD	cash before delivery
CD	customs declaration
C&D	collection and delivery
Cge Pd	carriage paid
Cert	certificate
C&F	cost and freight
Cert inv	certified invoice
C/I	certificate of insurance
CIF	cost insurance and freight. Import statistics are usually counted CIF because the total cost to the importer includes insurance and freight charges. For trade purposes it is important to be clear whether goods are sold CIF, C&F or FOB
CIF, C&I	CIF plus commission and interest
CIFANE	CIF and slight difference in exchange
CN	credit note, consignment note, cover note
CO	certificate of origin
COD	cash on delivery
CWO	cash with order
Cy	currency
D/A	document against acceptance, deposit account, days after acceptance
D/O	delivery order
D/P	documents against payment
D/W	dock warrant, dead weight
Dely or D/Y	delivery
E&OE	errors and omissions excepted
Ex	out of, after, without
FCL	full container load
FOB	free on board

IOU	I owe you
I/L	import licence
LC	letter of credit
mv	motor vessel
N/A	no account, not acceptable, not applicable
OC	open cover
OP	open policy
OR	owner's risk
pa	per annum
P/C	price current, per cent
pp	per pro, by power of authority, on behalf of
Ro-ro	roll on, roll off
SB	short bill, payable on demand or sight
SD	sight draft
SRCC	strikes, riots and civil commotion
TT	telegraphic transfer

Chapter 10
Privatisation and the Fish Industry

What is privatisation?

Privatisation is the transfer of state-owned assets from the public sector to the private sector (Vuylsteke, 1988). Generally, when we think of privatisation we assume it is a transfer of title to assets, but milder forms are also possible. Leases by the private sector of publicly owned assets may sometimes be a more desirable option for the transfer. Management contracts are another way in which the private sector can be brought into the management of state-owned businesses.

One of the dramatic changes that has occurred in the world of business and economics is the flood of privatisations that have taken place across the world. State-owned fish distribution systems, such as the Polish state-owned fish distribution company, fishing vessels, like the Polish Baltic fleet, fishing companies, such as those in most eastern European countries, harbours and ports, such as those owned by Associated British Ports in the UK, and eastern European fish farms have all been privatised. Many private investors have made much money from privatisation. Once again, to find out how businesses have made money the categories of architecture, reputation, innovation and strategic assets need to be brought into the analysis. Private investors have done especially well when they have recognised that the state-owned business has strategic assets which can be put to more profitable use under private ownership. Docks and harbours, fishing licences, reservoirs and water delivery systems come to mind. Also some private investors have taken over state-owned companies operating in very competitive markets and have been able to change the architecture – internally and externally – and to innovate, to transform their commercial performance.

So why does a country decide to privatise its industries?

(1) To reorganise its state-owned enterprises (state-owned businesses) to fit the country's current policies

A conviction has swept the globe that the state should not, if at all possible, be involved in running industry. This is because evidence has grown that privately owned businesses are usually significantly more efficient than state-owned businesses. This implies that they are able to produce their goods and services at lower unit cost than a state-owned company in the same line of business. When that experience is spread across an entire economy the higher cost of production of state-owned businesses compared to the private sector implies a smaller, or even negative, investible surplus and therefore a slower rate of economic growth. This in turn translates into long-term damage to living standards. The vast improvement in the economic performance of Poland following widespread privatisation is living proof of this proposition.

(2) To reduce the cost to the country's budget
An implication of industrial inefficiency is often that the businesses concerned are in receipt of government subsidies. It has been a common experience that this is the only way to keep some of them going. The consequence of an increasing budget deficit is that it has to be met by higher taxes, borrowing, or sometimes printing money. Whatever the outcome economists agree that it is undesirable for a country to have persistent structural deficits in its budget.

(3) To improve efficiency
Some of the expected efficiency consequences of privatisation are noted under (1) above. However, there is a further argument. It is that private business is generally competitive and competition is also a spur to increased efficiency. In order to capture markets, businesses are all the time on the look out for lower-cost means of producing their outputs.

(4) To create a wider distribution of ownership
A different reason for privatisation, which has been applied in a number of countries, is the conviction that a wider distribution of company ownership, in the form of shareholdings, is desirable. The theory is that people who own shares will feel that they have a stake in society and perhaps therefore be more responsible citizens, and also more contented people.

Main influences on privatisation policy

Government objectives

The type of privatisation selected depends on the policies of the government. For example, wider share ownership will point in one direction, raising as much money for the state in another, and disposing of the state-owned enterprises as quickly as possible (distress sales) is a third. Eastern European fish industry privatisations have been influenced by the desire of the states to transform their economies from state traders to liberal market economies.

The condition of the business

It will become evident when we examine state-owned companies that the condition of the business has a bearing on the method of privatisation selected. There are many different variables to consider, such as short- and long-term profitability, cash status, ownership of assets, efficiency of the workforce, etc. It was possible for Poland to sell its Baltic fleet to the fishermen because the vessels were earning cash surpluses, implying that at least the vessels had some value, albeit minimal. However, there are some eastern European fish industry privatisations where the best outcome is a single, possibly foreign, buyer who will change the employment of the assets from fish production. This might imply converting land currently used for fish farming to some other purpose. It might mean selling fishing vessels for scrap.

What is happening in the sector?

The state of the sector as a whole is also an important influence on privatisation policy. The privatisation of a fishing company under conditions where the outlook for access to high-valued resources is good is quite different from the privatisation of a carp farm faced with a declining market, and different again from the privatisation of a long-distance fishing company faced with the prospect of being excluded from its traditional fishing grounds.

Performance of the country's economy

The economy of the country is another influence on the method of privatisation selected. For example, a country with an active stock exchange opens up the possibility for stock market flotation, but in many developing countries there is little or no tradition of trading in stocks and shares and little liquidity (freely available cash or assets which can easily be converted into cash) to support it. Attitudes to employment may also be influential because the new owners may be prevented from making the business more efficient because they will not be allowed to implement labour cuts.

Nature of capital market regulation

In developed countries, capital markets probably face a reducing burden of regulation. Investors can borrow in one country and invest in another with little legal constraint. However, the flows of capital are still more restricted elsewhere, with rules about local partners, repatriation of profits, reporting procedures, taxation and other factors. In an unregulated market, privatisations can be marketed internationally, and this should yield a better price. Where there are restrictions the approach to potential buyers, and the expected price for the assets, will be affected.

Conclusions

Thus, obstacles to privatisation include:

- Financial condition and excessive liabilities of many state-owned businesses
- Lack of or weak financial markets

- Impact on employment
- Excessive debt of some state-owned businesses.

Planning and management of privatisation

Every privatisation needs to be carefully planned and managed. A privatisation is a one-off concentration of resources to bring about a disposal of state assets. It is, therefore, a project and the process needs very careful projectisation. Central governments wishing to engage in a privatisation programme need to consider the following issues.

Centralisation of decision making

Privatisation experts (such as Vuylsteke, 1988, and Frydman *et al.*, 1993) believe that there is significant advantage if the policy-making drive for transferring the ownership of assets comes from central government. Sectoral departments, such as a fisheries department, may be too bound up in their own industries to be able to take the strategic overview that is required. Moreover, when policy is initiated at the level of central government it certainly has more force behind it.

Transparency

Transparency implies that the public and its representatives understand precisely what is going on. The process of privatisation will be weakened if there are suggestions that some interests have been given an unfair advantage. Moreover, if there is corruption the process may be subject to later legal challenge. Key areas where transparency is important include: the valuation of assets, recognising that this may be difficult but nevertheless using common rules; the procedures adopted for implementation; ensuring that the bidding process is competitive and fair.

Projectisation of the process

Privatisation policy and strategy is determined at central government level. At departmental level the process should be organised into projects, with clear project objectives, milestone plans, bar charts, risk analysis and monitoring procedures.

Methods of privatisation

Public offering of shares

The government sells shares in the new enterprise to the public. The shares may be existing *stock* or newly created stock. The services of an *investment bank* (or merchant bank) as adviser are invariably required because government departments are unlikely to have the expertise necessary to complete the project.

The investment bank organises the preparation of a *prospectus*, which sets out the details of the privatisation proposal. This document must be prepared with legal advice to ensure strict conformity with the law. Normally it is also subject to other reviews so that it is readily understandable for the share-buying public. Also, the printed version of the prospectus should be produced to a very high standard.

A bank or a syndicate of financial institutions may *underwrite* the share-offering, guaranteeing the sale at some minimum price in case the public decides not to buy. There is a cost to the syndicate because they may have to pay for something above the going price, but there is also the benefit of receiving an underwriting fee. In any event, on the rare occasions in the UK that the public has not raised enough money for the shares on offer, underwriters have normally been able to sell the shares successfully at a later date to the public, and more especially to pension funds and insurance companies. For public offerings the normal procedure in the UK is to fix the price before the offer is opened. This has been somewhat controversial in the UK because government advisers have suggested prices that appear to be quite low. The offerings have been organised to ensure their success in terms of achieving sales of shares rather than maximising government receipts.

Shares may be offered nationally or internationally. Sometimes there may be incentives for employee participation, giving them first choice of a discount for shares. Sometimes restrictions on size of purchases or individual shareholdings or purchases by foreigners may be imposed. Sometimes it may be necessary to turn a public enterprise into a limited liability company before it can be sold.

Public offerings require that:

- the enterprise be a sizeable going concern with a reasonable earnings record or potential. The public have to have reasonable expectations that the business will be profitable in the future if they are going to be willing to subscribe. This could be the case for some fish businesses, but much depends on secure access to the strategic asset of managed fish resources.
- information about financial management is available. This can be a constraint in some countries where validated or audited accounting records are not available, or the accounting methodology does not meet the requirements of transparency. Eastern European accounting practices can be an obstacle to privatisation.
- there is liquidity in the local market. In many developing countries this may be a problem.
- the equity market be developed or at least there is some structured mechanism to reach, inform and attract the investing public. This has been relatively easy in the UK where the press has been the main means of informing the public. Other advertising media, including television, have also been used.

Private sale of shares

The government identifies a purchaser who is willing to buy either all or part of the shares in a state-owned company. Thus, there is direct acquisition of state-owned shares by another corporate entity or targeted group. The process implies that a limited liability company, of which the government owns the shares, has already been created because there has to be a company for there to be shares in it. This may often not be the case and the company will need to be established first. This method of privatisation is most suitable for cases when a business has good potential which is not currently being exploited. Private sales may also be appropriate when equity markets are weak or when there is no other mechanism for reaching the investing public.

Options for implementation include the following:

- A fully competitive procedure in which bids are invited for the shares; in order to remove frivolous buyers from the process a charge for the bidding documents may be made.
- Sometimes governments may choose to narrow down the field with an initial sift of potential buyers so that only a small number make a full bid. This is known as prequalification of bidders.
- Governments may opt to enter into direct negotiations with potential buyers. The problem here is that the transparency requirement will soon be breached because any buyer will want to keep matters confidential.

If a business is to be sold either in whole or in part, some alleviation of debts may be required. The new owners may make some of the workforce redundant as they introduce measures to improve efficiency.

Sale of government-owned assets

Under this procedure the government's assets are sold to the highest bidder, perhaps through an auction or through direct negotiation. This method is most suitable when the business is not a going concern and the government wishes to put the assets of the business into the hands of a company or individual who can make better use of them. What to do about existing liabilities is a problem – perhaps more so in this case because it is not easy to see how such liabilities could be attached to the state-owned assets if there is no business with a legal personality to own them. As with most privatisations, employment issues may arise because the new owners may want to restructure the business. An example of an eastern European country (Bulgaria) is noted below in which the state-owned assets have not been sold to the highest bidder, but actually are in the process of being handed back to the former owners before the land was appropriated by the state.

Reorganisation of the state-owned business into its component parts

The government may opt to split up its business into its components. For example, some of the larger collective entities in eastern Europe may attract a better market price if they are broken up. Fish farming in Romania is typically organised into large-scale fish societies (see Appendix to Chapter 7) so breaking these up and selling of the pieces for a variety of different uses may be a logical solution. Different methods of privatisation may be suitable for different parts of the business with the added advantage that competition will be created.

New private investment in the state-owned business

An example of private investment in a state-owned business in the UK is that of BP. Over the years the British government has encouraged the dilution of its own equity holding to the point where the company is now privately owned. This has been done through the periodic sale of shares to the public and institutions.

Management/employee buy-out

Some management/employee buy-outs (MBOs) work, and the employees who were fortunate enough to participate in the bid become owners and then prosper. Some MBOs fail with unfortunate consequences for the buyers. Frequently employees have to borrow to purchase the company, and this is why these buy-outs are often referred to as being *leveraged*. It is, of course, disastrous for employees if they have used their personal assets as collateral for the borrowings and then the business fails.

In the case of a leveraged buy-out the gearing ratio (the ratio of borrowing to equity) can be very high, as much as 5, which is much higher than the usual standard level of 1. This implies that the company has to be consistently profitable because it must meet its interest payments. If it is successful, and the interest payments are easily met, then the return to equity can be very high indeed because the new owners will have put relatively little of their own money into it but achieved a very high return.

Leases and management contracts

If governments want to introduce the private sector into the state-owned sector then there are other things they can do. It is possible, for example, to lease state-owned assets to private business. Or management contracts can be agreed to take over on a temporary basis assets belonging to the state. Very difficult issues of incentives for the management group can arise in these cases because of the temptation on their part to take a short-term view in their own interests rather than the longer-term view in the interests of the state-owned enterprise.

Bulgaria: an example

Bulgaria has been one of the slowest of the old eastern European state trading countries to enter the market economy. However, it now has a government that is committed to selling state assets. The pressure for this is partly international as one of the conditions for international support. Bulgaria has a privatisation law, the Transformation and Privatisation of State and Municipal Enterprises Act, April 1992 (Zlatanova and Kissiov, 1996). It also has a Privatisation Agency as a result of this. A commitment was made in 1991 to the International Monetary Fund (IMF) and the World Bank to initiate rapid privatisation. Unfortunately little has been done mainly because of protracted political arguments. Indeed, it has been argued that many of the country's current problems stem from a failure to submit industry and agriculture to market forces soon enough.

The Act covers the sale of state and municipal companies following their transformation into joint-stock or sole owner limited liability companies (see Chapter 6). The privatisation process falls into the second category noted above – the private sale of shares. A number of fish farm societies have now been transformed into sole owner companies although they have not yet been sold to the private sector. The Act also provides for the preparation of comprehensive annual privatisation plans to be prepared by the Privatisation Agency.

The process of land restitution to former owners began in 1991, when the National Assembly, still dominated by the Bulgarian Socialist Party, passed the Act on Ownership and Use of Farmlands. This provided for the restitution to the former owners and their heirs of the land forcibly collectivised in the 1950s. Not surprisingly progress has been very slow, not least because it has proved very difficult to establish who the actual landowners were. At the present time, although much land has, officially, been privatised legal title to land has not been obtained.

Privatisation and the fisheries sector

Fish businesses are generally small (compared with the major enterprises in an economy). It seems appropriate then that many fisheries state-owned businesses should be sold to single buyers, either before or after their transformation into limited liability companies.

State of the business

Some examples follow. The fish market in Vanuatu has recently been sold to a private buyer (see Chapter 16). It was a limited liability company and a new buyer will make use of the assets. It was losing money and, no doubt, the new owner believes that the assets can be used to greater effect, perhaps making use of the facilities for the wholesaling and retailing of other products as well as fish. The new owners are likely to have in mind some new configuration of buying and selling which will restore the profitability of the cold storage, cool storage, freezing facilities and sales area. Thus, it is through a change in the architecture that the new owners hope to make profitable use of the assets.

The Polish Baltic fleet has been privatised for some years. The vessels were transferred to their skippers who pay for them through a small deduction from the money received for fish. It must be noted in this case that these old cutters would have had very little value on the open market so this was probably a good solution. From the point of view of the new owners they have received assets at little cost and, crucially, enjoy the strategic asset of access to Poland's Baltic fisheries resources.

Bulgarian and Romanian fish farms are still at the first stages of privatisation in legal terms, although in practice they receive no money from central government. They are, in general, not in the kind of condition which would attract major buyers. For them to be worthwhile to private owners there should be some way in which the new proprietors can visualise making more productive use of the assets than at present.

The western European fleets have always been privately owned, but the ports have not. Associated British Ports (ABP) owns some fishing ports, including Hull and Grimsby. The privatisation was a great success precisely because ABP has been able to utilise the assets of the commercial dock facilities so much more efficiently since privatisation. The docks were strategic assets which were seriously underexploited.

The Kamchatka Peninsular, at the extreme east of Russia, lies between the Bering Sea and the Sea of Okhotsk. The Kamchatka Trawling and Refrigeration Fish Company was an early Russian privatisation. The company was advised by the management consultancy company McKinsey and Co. The company has cut capacity back to a core of 32 efficient ships and has stopped wasting money repairing redundant fishing vessels. It is now partially managed by Icelanders and exports filleted fish – Alaskan pollack – crab and salmon and is reckoned to be a great success.

Targeting ownership

Once a government has decided to privatise a state company, it needs to decide upon the kind of owner it is looking for. In the case of small businesses, the dominant fisheries sector form, it can advertise nationally or internationally. A number of small fish businesses in eastern Europe have been sold to their managers (the MBO). Large businesses require the expert advice and planning skills of an investment bank.

Valuation and pricing of shares

In principle, the value of a business is the net present value of the assets at the time of privatisation (see Chapter 15). In practice, this is almost always impossible to determine with any accuracy because of uncertainty about the future. Much depends on the quality of the future management and to what profitable use it can put the assets. A crucial issue for fish catching businesses is secure access to fish resources for as low a licence fee as possible. The access represents an important strategic asset which can be very helpful for long-term profitability. Again, expert advice is needed to make the necessary judgement.

Absence of developed equity markets

In some developing countries and some of eastern Europe it will often be the case that the equity markets are thin or non-existent. So, finding a direct buyer might be the only realistic solution for privatised fish businesses. In many countries, such as those of Western Europe, equity markets are well-developed so the flotation route for large companies will be realistic. Private buyers are more appropriate for smaller businesses.

Summary

Privatisation is the transfer of state-owned assets from the public sector to the private sector. Generally, when we think of privatisation we assume it is a transfer of title to assets, but milder forms are also possible. Leases by the private sector of publicly owned assets may sometimes be a more desirable option for the transfer. Management contracts are another way in which the private sector can be brought into the management of state-owned businesses.

Main influences on privatisation policy

- Government objectives
- The condition of the business
- What is happening in the sector
- Performance of the country's economy
- Nature of capital market regulation.

Obstacles to privatisation

- Financial condition and excessive liabilities of many state-owned businesses
- Lack of or weak financial markets
- Impact on employment
- Excessive debt of some state-owned enterprises.

Planning and management of privatisation

Every privatisation needs to be carefully planned and managed. A privatisation is a one-off concentration of resources to bring about a disposal of state assets. It is, therefore, a project and the process needs very careful projectisation.

Methods of privatisation

- Public offering of shares
- Private sale of shares
- Sale of government-owned assets
- Reorganisation of the state-owned business into its component parts
- New private investment in the state-owned business
- Management/employee buy-out
- Leases and management contracts.

Privatisation and the fisheries sector

Fish businesses are generally small (compared with the major enterprises in an economy). It seems appropriate then that many fisheries state-owned businesses should be sold to single buyers, either before or after their transformation into limited liability companies.

Chapter 11

The Common Fisheries Policy of the European Union

The Common Fisheries Policy (CFP) of the European Union is a very interesting phenomenon for those brought up in the Anglo-Saxon world of relatively little regulation. In microcosm it demonstrates the differences between how things happen under the EU and how things used to happen in Britain. A massive structure of law and regulation has been erected to govern the way fishing takes place. It is a structure that is always changing. Businesses in the industry need to make up their minds how they are going to stay on top of all the detail.

Trade associations and fish producers' organisations are probably the most suitable vehicles for maintaining a watching brief and keeping the industry, including aquaculture, alert to what is going on. It will be obvious from this chapter that the CFP presents considerable opportunities for business. These arise through subsidies, such as grant assistance of various kinds, and through the creation of what has been termed in this book as strategic assets, such as fishing licences. It also threatens fish businesses if they are not alert to its development. For example, fishing grounds may be lost to them, and hygiene and environmental regulations may put processors or fish farmers out of business. Entrepreneurs need to calculate carefully the best balance between cooperation and competition in this situation to gain the best advantage. It also needs to be emphasised that the European Commission is very good at keeping the world informed of developments.

The CFP: an application of the Treaty of Rome or a conspiracy?

The general extension of fishing limits to 200 miles or thereabouts took place during the 1970s. Iceland's proactive and unilateral defence of its fishing interests ran counter to the interests of the distant-water fishing. The result was no less than three cod wars (the defence of British trawlers with tugs and warships) in the waters around Iceland (1958, 1972 and 1975). The motive for the third confrontation was Iceland's extension of its fishing limits to 200 miles – a pattern which had by then become the international norm. In the same decade the British fleet suffered progressive loss of access to the Barents Sea (north of Norway), the Faeroe Islands and the North West Atlantic. It is easy, therefore, to interpret the

CFP as a conspiracy cobbled together by the six original members of the EU in 1970 to ensure that Britain, and the other 1973 new members (Ireland and Denmark) did not follow Iceland and other maritime states in restricting access to fishing grounds. The 1970 policy expanded the *acquis communautaire* over the fisheries sector which was already defined as being part of agriculture under the Treaty of Rome (1957), a process which has continued since then. A summary of the evolution of the CFP is given in Fig. 11.1. But whatever the origins of the CFP it

The four main components of the CFP	(1) Markets (2) Structures (including FIFG) (3) Access and conservation (4) External relations
Acquis communautaire	The principle means that new members must accept existing EU legislation. Transitional arrangements can be negotiated (known as derogations)
1973 Denmark, Ireland and the UK joined	These states obtained a derogation from Regulation 2141/70 for 10 years. This gave vessels exclusive use of a 6-mile limit and partial 6- to 12-mile limit
Equal access	This principle has been enshrined in the CFP from the beginning. In 1997 the UNCLOS Agreement was signed, but 200-mile EEZs of the EU are shared
Third country agreements	Undertaken by the EU on behalf of member countries to retain access threatened by 200-mile EEZs
Regulation 170/83	Extended 6- and 6- to 12-mile derogations
Relative stability	Refers to allocations of TACs to member states for each stock
Shetland Box	Correct name 'North of Scotland Box'. Restricted access. Officially ends in 2002, but likely to continue
Basic Regulation 3760/92	Replaced 170/83
Permit System for fishing vessels	Introduced in the Treaty of Corfu, 1994, for accession of Austria, Finland and Sweden. All vessels now controlled by Brussels through member states
Multi-Annual Guidance Programme (MGP)	MGPs I-IV (I, 1983–1986; II, 1987–1991; III, 1992–1996; IV, 1997–2001). Aimed at reducing size of the fleet

FIFG, Financial Instrument for Fisheries Guidance; UNCLOS, United Nations Conference on the Law of the Sea; EEZs, exclusive economic zones; TACs, Total Allowable Catches.

Fig. 11.1 Summary of the evolution of the Common Fisheries Policy.

is hard to believe that the political clock can be turned back to the 1960s, and national control reasserted over the sector.

The EU industry

The 300 000 fishermen of the EU catch around 7 million tonnes of fish a year, which is about 7% of the world total. Of this, the UK and France each catch about 800 000 tonnes. Spanish fishermen are responsible for 1.4 million tonnes and the Danish catch is 1.8 million tonnes but falling fast as controls on fishing with very small mesh nets for fish to reduce to fish meal are extended. The EU is the largest single market for fish in the world. It relies on imports from the rest of the world to meet a large part of its requirements.

The UK component

The catching and processing of fish and aquaculture account for less than 1% of the UK gross domestic product (the value of production within the geographical boundaries of the UK) and directly employ less than 1% of the working population. However, these activities and others related to them are important to many areas around the UK coastline where they often account for much more of the local or regional employment and contribute to the local tourist industry. Furthermore, in many cases fishing vessels, fish processing and aquaculture facilities are located in areas with unemployment above the national average, where economic activity needs development.

The Ministry of Agriculture, Fisheries and Food and its Scottish counterpart publish comprehensive information about the UK industry (see Chapter 7 for data sources). The UK catching sector employed about 22 000 fishermen in 1994, and numbers are in decline as the fleet size falls. About the same number is recorded by MAFF as being employed in processing and more are employed in associated industries such as shipbuilding and repairs and equipment manufacture. At the retail level there are about 3000 fishmongers employing 7000 people and fish is also sold through supermarkets. Fish is also sold through many other outlets, such as fish and chip shops, restaurants and the institutional market (such as prisons, hospitals and schools).

There are about 11 000 vessels in the UK fishing fleet. Of these, about 8000 are 10 m or less in overall length and catch 5% of the catch. Since 1990 the fleet has declined slightly. In Scotland the fishing industry continues to be significant and is a major source of employment and income in the more remote parts of the country. The significance of the fishing industry in Scotland is illustrated by the fact that whilst the Scottish population accounts for only 9% of the UK population, fish landings in Scotland accounted for over 60% by weight and about 60% by value of all fish landed in the UK. The Scottish aquaculture sub-sector is also significant for salmon and trout.

Current policy

The fully fledged CFP was adopted by the EU in 1983. It included some bias towards regions that are dependent on fishing, as well as compensating those countries that had suffered losses of fishing grounds in third country waters. Greenland, which would have been an important source of fishing grounds for the long-distance fleets, left the EU in 1985. The accession of Spain and Portugal in 1986 presented a considerable challenge for the CFP, because Spain still had a large fishing industry and prioritised access to the fishing grounds of other member states in its negotiations. The Commission's Directorate-General for Fisheries publishes a very informative Information File on the developing CFP which explains the policy in much more detail than here (Commission of the European Communities, 1996).

Access to fishing grounds

The equal access principle that has been in place since 1970 is in practice not universally applied. Coastal waters, from 0 to 6 miles, are, with minor and defined exceptions, reserved for small-scale fishermen. In some cases this privileged status extends to 12 miles. Other cases of limited access also apply, for example, the Shetland Box restricts the number of vessels over 26 m permitted to fish within a defined area of the North Sea.

UK vessels have access to Spanish and Portuguese fishing grounds, but with minor exceptions British fishermen have shown little interest in these. This is because the rich, relatively shallow waters are in the North Sea, the western approaches to Britain and Ireland and the north and west coasts of Scotland. However, British fishermen often operate in waters on the east side of the North Sea so to that extent take advantage of the principle of equal access.

Conservation measures

The main instrument of fisheries management is the Total Allowable Catch (TAC). The Council of Ministers decides the TACs following scientific advice. They are then divided between member states into national quotas on the principle of relative stability. The EU also requires fishing vessels to be licensed. There are also technical regulations governing matters, such as permitted mesh sizes, the minimum permitted size of fish and gear restrictions. The conservation policy is managed by member states, who operate the fishery protection vessels and aircraft, and is held up as an example of subsidiarity.

The International Council for the Exploration of the Sea (ICES), the Northwest Atlantic Fisheries Organization (NAFO) and the International Commission for the Conservation of Atlantic Tunas (ICCAT) provide the scientific advice upon which the TACs are determined. There are analytical TACs, based on scientific advice, and precautionary TACs where scientific data are insufficient.

Marketing

The CFP's marketing policy has four main instruments:

(1) The application of common standards for marketing products
(2) The recognition of Fish Producers' Organisations (FPOs). There are now about 150 officially recognised FPOs within the EU
(3) The price regime, consisting of a withdrawal price system and some other compensatory mechanisms
(4) Trading arrangements with third countries – to ensure that the market requirements of the EU are met without damage to EU producer interests.

In the fishing industry there is widespread recognition that the system has, in general, worked well. In particular, the innovation of FPOs has been a useful contribution.

Structure of the industry

The CFP includes provisions to encourage the development of a modern competitive industry and, crucially, to remove excess capacity. Financial assistance is available through the Financial Instrument for Fisheries Guidance (FIFG) and the Pesca initiative which support diversification of the industry.

Third countries

Many EU vessels fish in international waters. Spanish vessels operating in Moroccan waters are particularly significant (about 730 vessels), so the growing desire of Morocco to exploit its resources with its own vessels imposes considerable strain on the Spanish industry. EU vessels fish in the waters of third countries on the basis of two types of agreement:

(1) A reciprocal agreement (Norway, Iceland, the Faeroe Islands and the Baltic States), which is an exchange of fishing opportunities for EU fleets and the fleets of these countries.
(2) Fish for ecus. Under these, payments are made to the countries concerned to give access to EU vessels. The payments are made to governments and include grants for studying and training.

   The agreement with Morocco includes a section on the long-term development of the country's industry. The agreement with Argentina is based on setting up joint-venture consortia. The EU has recently commissioned a major research study to examine the costs and benefits of such agreements.

The Pesca initiative

Regional policy

The Pesca initiative arises from the impact of the decline of the catching sector, but is integrated into other EU policies, particularly regional policy. Thus, an explanation of Pesca requires a small diversion into regional policies.

Numerous regional concepts are used in the UK. Some examples are Planning Regions (such as Yorkshire and Humber), local authority boundaries, Travel to Work Areas, Rural Development Areas, Enterprise Zones, Assisted Regions (which subdivide into Development Areas and Intermediate Areas) and Unassisted Areas. A variety of regional definitions have been adopted by national governments over the years for different administrative purposes. For example, the concept of the Travel to Work Area is used extensively by local authorities in the UK because it coincides well with the economic area associated with each main centre of population The concept was also used in the UK components of a series of socio-economic studies of the fishing industry commissioned by the EU in 1991 (Report 1991 from the Commission to the Council and the European Parliament on the Common Fisheries Policy; Commission of the European Communities, 1991, p.49).

The EU has long recognised the concept of the region. The European Regional Development Fund was established in 1975 in recognition of the fact that there are disparities of income and employment between various regions in the EU. The UK has been a beneficiary of this fund. Following the adoption of the Single European Act in 1986 the EU increased the resources available for regional policy. Five priority objectives were defined by the Council of Ministers for the structural funds (European Regional Development Fund, European Social Fund and the European Agricultural Guidance and Guarantee Fund). Of these priority objectives, Objectives 1, 2, 5(a) and 5(b) are specifically regional. These may be summarised as follows:

- *Objective 1:* promoting the development and structural adjustment of the regions whose development is lagging behind
- *Objective 2:* converting the regions seriously affected by industrial decline
- *Objective 5(a):* accelerating the adjustment of agriculture and fisheries
- *Objective 5(b):* promoting the development of rural areas.

The concept of the region is recognised in the Maastricht Treaty (Article 198A). It seems likely that subsidiarity, or the principle that decisions should be taken at the lowest possible level, will be achieved, at least in part, through regional policy. The Maastricht Treaty established a 'Committee of the Regions' with advisory status. The UK, with France, Germany and Italy, has 24 members on this Committee. Other EU member states will have lower numbers, reflecting their smaller populations (Belgium 12, Denmark 9, Greece 12, Spain 21, Ireland 9, the

Netherlands 12, Luxembourg 6 and Portugal 12). The Committee of the Regions has similar status to the Economic and Social Committee and it will be able to develop its own initiatives (Article 198C of the Maastricht Treaty).

Pesca

The provisions for providing financial assistance for the structural development of the fishing industry were integrated into the general regional development funds (Council Regulation 2080/93). The Financial Instrument for Fisheries Guidance (FIFG) was created, subject to the general principles of the Structural Funds which are allocated in accordance with the regional development objectives (for details see Commission of the European Communities, 1995a,b).

The Council of Ministers decided that the socio-economic impact of restructuring the fishing industry would be a criterion of eligibility for access to the Structural Funds. Thus, provided the workforce and enterprises of the fishing industry are within the relevant regions, they are able to gain access to the Regional Fund and the European Social Fund as well as the FIFG money.

Information about the Pesca initiative is well communicated in the publication *Pesca Info* (Commission of the European Communities). The Pesca initiative combines several elements and has the following objectives:

- To put the fishing industry into a condition where it can succeed in its own transformation, by supporting the sector in association with, but not replacing, the actions under Objective 5(a).
- To help it cope with the social and economic consequences, by providing aid for the redeployment of the workforce and diversification of the enterprises in the industry.
- To contribute to diversification in the coastal areas concerned, by creating employment.

The EU contribution to Pesca is 251.5 million ecus at 1994 prices for the period 1994–1999, half of which is for Objective 1 regions (underdeveloped areas). In addition to providing financial support for restructuring the sector, the Commission intends to use Pesca to promote dialogue and other collective action between the various political, administrative, scientific and economic institutions so as to improve relations between competing trades, communities and activities at a local, regional, national and Community level.

Areas where bids are invited include:

- Training of fishermen and tradesmen
- Networking of training institutions
- Cooperation in matters relating to training programmes
- Joint management of fisheries
- Initiatives aimed at reducing conflict

- Transregional initiatives for the marketing of products
- Projects concerned with marine safety
- Economic diversification of areas dependent on fisheries
- Conversion programmes
- Promotion of services aimed at reducing the isolation of remote areas
- Meeting of interregional partnerships between areas dependent on fisheries
- Innovative job-creating initiatives
- Cooperation and networking in fisheries jobs
- Local employment initiatives
- Initiatives in favour of women.

Cooperation and networks concern professional organisations, training institutes, local or regional authorities, associations or individual companies. Projects should lead to concrete measures.

Examples of Eligible Projects

- Diversification of activities
  - exploitation of quayside auctions and fishing ports as tourist attractions;
  - a combination of tourism and fishing;
- Improving the professional qualifications of sea fishermen
  - improving the content of the initial training, including some commercial training, as, at present, navigators are in receipt of training, but managers are not;
  - training in the biological and economic management of fisheries and training in the CFP;
  - general training in order to improve the status of sea fishermen and, if necessary, to equip them to do another job;
- General projects
  - mapping, medical, seasonal adjustment of supply, fisheries management, conflict resolution, marketing, product development, pilot measures for the development of quality labels.

Eligible areas in the UK

Areas in the UK defined as being dependent on the fishing industry, and therefore eligible for Pesca support, include the following:

- *England and Wales (county and town)*
  Northumberland: Amble, Blyth
  Durham: Hartlepool
  Tyne and Wear: North Shields, Sunderland
  Cumbria: Whitehaven
  East Yorkshire: Hull, Bridlington

Devon: Plymouth, Brixham
Cornwall: Padstow, Falmouth, Newlyn
North Yorkshire: Scarborough, Whitby
Suffolk: Lowestoft
Lancashire: Fleetwood
Dorset: Weymouth
East Sussex: Hastings, Newhaven
Essex: Leigh-on-sea
Norfolk: King's Lynn
Lincolnshire: Grimsby, Boston
Gwynedd: Holyhead, Pwllwli, Portmadoy, Conwy, Amlwch, Aberdovey
Dyfed: Aberystwyth, Fishguard, Milford Haven, Burry Port, Cardigan and Carmarthen Bay
West Glamorgan: Penclawdd
- *Scotland (region and town)*
Grampian: Macduff, Buckie, Lossiemouth, Fraserburgh, Peterhead, Aberdeen
Tayside/Fife: Arbroath, Pittenweem
Border: Eyemouth
South West: Ayr
Highlands and Islands: Argyll, Caithness and Sutherland, Lochaber, Orkney, Wester Ross: Shetland, Skye and Lochalsh, Western Isles
- *Northern Ireland (district and town)*
Ards and Down: Portavogie, Ardglass
Newry and Mourne: Kilkeel, Annalong

Main issues

Quota hopping

Countries are allocated national quotas, their shares of the TACs. However, there is free movement of capital in the EU, so any EU citizen can buy a licensed fishing vessels from anyone else, and if the licence attaching to that vessel entitles it to a share of a quota, it goes to the buyer. The previous Conservative government said it would veto any treaty unless this problem was resolved. Their proposal was for an arrangement whereby the owner of a UK licensed vessel would be obliged to operate from Britain. It is not easy to see how this would have solved anything because forcing an EU business to operate from Britain would simply be a minor irritant for potential buyers of British licences – the cost of a shell company and a brass plate. Moreover, for many years British-registered vessels have been owned wholly or partly by foreign companies and individuals and have operated in and out of continental ports, frequently attracted by the higher price of fish.

The Labour government will have to find a different solution – one which is consistent with the provisions of the Treaty of Rome, guaranteeing free movement of capital and freedom of establishment. A way forward might be an extension of the idea of restricted boxes, such as the Shetland Box, so that fishermen of certain regions where the industry is socially important might enjoy ring-fenced quotas and fishing areas.

Fleet reductions

The Commission has set targets for the reduction of the capacity of fishing fleets (Multi-Annual Guidance Programmes). Under the third Programme (1992–1996) countries were committed to a reduction of 20% in the capacity of the fleets targeting demersal species (i.e. white fish, such as cod and haddock), a target the UK has had difficulty in achieving. The fourth Programme was published at the beginning of 1998 (OJ L 39) and envisaged reductions of 20 to 30% in the capacity of the principal fleet segments. The 'quota hoppers' are still in the UK fleet, although owned by foreign companies, so they count towards UK capacity – which seems anomalous. One way out of the problem might be for the incentives to be increased, so that fishermen find it more attractive to decommission than to sell the vessels abroad.

Other effort controls

The TAC/national quota system will remain in place, supported by mesh size and other technical regulations. However, the EU is moving towards more controls on inputs, such as restrictions on the number of days vessels are permitted to fish. Since 1992 it has had powers to define areas where fishing is barred, restrict permitted exploitation rates, including days at sea, establish ceilings on catches, control gear and create incentives for selective fishing.

Expansion of the pelagic fleet

The world market for pelagic fish, such as herring and mackerel, has undergone dramatic change, probably in response to the opening up and rising prosperity of eastern European markets. Very powerful multipurpose vessels have been constructed to exploit the resources which are now under serious pressure. The licensing system in place has not been strong enough to moderate this development.

The Mediterranean

Large numbers of fishermen work in the Mediterranean and many of the fishing areas are shared with non-EU members. Some aspects of the CFP – marketing and structures – are already in place. However, serious fisheries management, involving cooperation between EU members as well as non-members, is still to come.

Enlargement

The fishing industries of eastern Europe are not, in general, large. Even so, the countries planning to join the EU need to ensure that the EU-type regimes for marketing, hygiene standards, marketing by producers' organisations, fisheries management and licensing are in place. For example, Poland has a significant Baltic fleet, and the rump of a long-distance fleet. Its Baltic fisheries need to be properly integrated into the EU system and it can reasonably expect the EU to negotiate on its behalf in distant waters.

International relations

The EU is moving from a system of simple payment for access for EU vessels to a partnership arrangement. The current agreement with Argentina, which involves long-term commercial links and research collaboration, as well as access for a restricted number of vessels, is held up as an example.

Socio-economic development

The number of people employed in the EU catching subsector is bound to decline from its present level of about 300 000 as fisheries management bites at fleet capacities. Developing alternatives for unemployed fishermen in those communities dependent on the industry, which are often remote (such as the west coast of Scotland) or have high levels of unemployment (such as Hull), is a problem of which the EU is aware, and programmes to deal with them, such as Pesca, are in place. However, making them work is another matter.

Monitoring the activities of the fishing industry

Council Regulation 2847/93 of October 1993 established a new monitoring system covering the whole CFP. The Regulation requires that all activities must be monitored, both at sea, to ensure compliance with technical measures, and on shore, where landings, sales, transport and storage regulations need to be followed. Inspectors from the member states must be in a position to collect comprehensive information to account for the origin of fish on every step of its journey until it reaches the customer's plate. Thus, log books, sales notes and travel documents are important. The EU is moving towards a system whereby catches are monitored using satellites. Pilot projects were conducted in member states between 1994 and 1995. A new regulation now allows for all vessels over 24 m long or at sea for over 24 hours to be fitted with automatic tracking devices. Log books are held, and include duration of fishing on particular grounds and how long fishing gear has been immersed. Catches are registered in real time and these data are followed up with records. When a quota is exhausted, fishing for it should cease. Council Regulation 847/96 of 6 May 1996 introduces flexibility. If a quota is not exhausted in 1 year it may be increased in the next, and vice versa.

Summary

The EU industry

The Spanish catch is about 1.4 million tonnes and the Danes catch 1.8 million tonnes, a figure that is in decline as controls on fishing for reduction to meal are extended. The EU is a very large market for fish and relies extensively on supplies from other countries to meet its requirements.

Current policy

The fully fledged CFP was adopted by the EU in 1983.

Access to fishing grounds

The equal access principle has been in place since 1970. There are derogations from this which are likely to continue into the new millennium.

Conservation measures

The main instrument of fisheries management is the total allowable catch (TAC). The Council of Ministers decides this following scientific advice. The TACs are then divided between member states into national quotas on the principle of relative stability. The EU requires fishing vessels to be licensed. There are also technical regulations governing matters such as permitted mesh sizes, the minimum permitted size of fish and gear restrictions.

Marketing

The CFP's marketing policy has four main instruments:

(1) The application of common standards for marketing products
(2) The recognition of fish producers' organisations (FPOs). There are now about 150 officially recognised FPOs within the EU
(3) The price regime, consisting of a withdrawal price system and some other compensatory mechanisms
(4) Trading arrangements with third countries – to ensure that the market requirements of the EU are met without damage to EU producer interests.

Structure of the industry

The CFP includes provisions to encourage the development of a modern competitive industry and, crucially, to remove excess capacity. Financial assistance is available through the Financial Instrument for Fisheries Guidance and the Pesca initiative which supports diversification of the industry.

Third countries

Many EU vessels fish in international waters. Spanish vessels operating in Moroccan waters are particularly significant.

The EC Pesca initiative

The Pesca initiative arises from the impact of the decline of the catching sector, but is integrated into other EU policies, particularly regional policy. Its objectives are:

- To put the fishing industry into a condition where it can succeed in its own transformation, by supporting the sector in association with, but not replacing, the actions under Objective 5(a).
- To help it cope with the social and economic consequences, by providing aid for the redeployment of the workforce and diversification of the enterprises in the industry.
- To contribute to diversification in the coastal areas concerned, by creating employment.

Main issues

- Quota hopping
- Fleet reductions
- Other effort controls
- Expansion of the pelagic fleet
- The Mediterranean
- Enlargement
- International relations
- Socio-economic development.

Monitoring the activities of the fishing industry

Council Regulation 2847/93 of October 1993 established a new monitoring system covering the whole CFP. The Regulation requires that all activities must be monitored, both at sea, to ensure compliance with technical measures, and on shore, where landings, sales, transport and storage regulations need to be followed.

The CFP is a reality which some fishing interests in the UK are continuing to resist whilst others, including successive British governments, have accepted. Fish businesses need to bear in mind Chapters 2 to 5 of this book when they are looking for commercial advantage in the system. The fundamental question is still that of obtaining the right mix between strategic cooperation and defection when dealing with the CFP.

Chapter 12
Setting up a Fish Auction

This chapter is included because the mechanism of setting up a fish auction is sometimes a puzzle for developing fisheries or indeed fisheries emerging from a state-owned fisheries sector. People are not always clear how auctions work. The chapter may also be useful for businessmen new to the fishing industry who require a brief appreciation of the fish auction system as currently operated in the EU. In terms of the theory outlined in Chapters 4 and 5, auctions are *games* which demand a degree of collaboration as well as competition from the *players*. Buyers and sellers need to agree that this is the best way of allocating landings and they all need to accept the principles by which the auction is conducted; for example, box lay-out, time of start, order of sales, grading principles, the signalling of offers and acceptance of offers, the determination of size of *lots*, and payment rules are all issues which require mutual acceptance. The fish auction is, therefore, part of the *architecture* of the fish business. The rules (also outlined at the end of Chapter 5) that the businessman should use to decide on his participation are exactly applicable to auctions:

(1) Understand the personalities and motives of the other players.
(2) Work out the expected gains from playing the game.
(3) Ask himself if he can change the rules of the game to his own advantage.
(4) Ask himself how the benefits of the game are going to be delivered.
(5) Decide on the appropriate tactics or strategies – which may be mixed.
(6) Decide on whether the scope of the game is susceptible to change, to his own advantage.

It is important to recognise that fish producers, acting through their producers' organisations or similar producer-run bodies, can sell their fish by four different methods: (1) individual vessel sales (presently possible but subject to FPO rules); (2) contract sales of the FPO as a whole; (3) sale by auction; and (4) private contracts. When dealing with the more homogeneous bulk products, such as herring or mackerel, it may suit the fish producers to sell under a contract.

Perhaps one of the most interesting questions for fish processors is whether or not to buy fish at fish auctions or to seek private deals with individual suppliers. The answer differs between fish merchants, depending on architecture. Some merchants have special requirements arising from the needs of their customers, which they believe are best achieved through private transactions – size, species and quality. Other merchants can more easily meet their requirements by buying through the auction system.

Establishment of fish producers' organisations

Fish producers' organisations (FPOs) are not necessarily a precondition for successful fish auctions, but they, or some entity equivalent to them, certainly smooth the path. FPOs are voluntary associations of fishing vessel owners and operators formed in order to enter joint arrangements for the production and sale of their fish, as defined under EU regulations. These associations are a successful component of the Common Fisheries Policy of the EU (see Chapter 11). Within them the fish catching sector has been able to exercise more control over the conditions for the sale of fish and in some cases actually takes over the role of the fish selling agency. This is what was intended by the EU legislation. Also, importantly, in many places they have ensured that fish selling takes place in an orderly and transparent manner under the ultimate authority of the fish producers.

In Grimsby, for example, decisions about landing and marketing rules are now taken by the Board of the Grimsby Fish Producers' Organisation, and not, as in the past, by the Grimsby Fishing Vessel Owners' Association. A recent study of Baltic fisheries (Directorate-General for fisheries (1995) *Baltic Fisheries – An Integrated View*) notes the formation of Fish Producers' Organisations in the former German Democratic Republic and also in Sweden, which are now helping to ensure the orderly and transparent sale of fish, although it seems that auctioning as a sales method as such has not yet taken hold in the Baltic region.

Location of auctions

With the fish catching sector in charge, the fishermen can now decide, through their FPO, where they want the fish auctions to take place as well as how they want them run. It is not necessarily obvious that they should be undertaken in ports close to the fishing grounds. The evidence is that a more important criterion is the availability of buyers who are willing to enter into competitive bidding under the auction format. Hull and Grimsby both provide examples of locations where direct landings are limited, but lively auctions continue to flourish. On the other hand, there are many examples of fishing ports and harbours where the fishermen opt to send their fish overland or by some other means of transport for sale by auction elsewhere, such as Icelanders or Scots sending their fish to the Humber markets. So the best advice is to look for a place where buyers can reasonably be expected to attend. Auctions have to be set up in collaboration with the port owners and operators and also in collaboration with the fish merchants (the buyers).

An alternative to the option of the FPO taking the decision is the establishment of auctions by independent commercial competing fish selling companies wherever investors choose to set one up. In many parts of the world, fish selling by auction is a commercial service offered to fishermen for a commission. The company has to win the support of fishermen who will then judge whether their sales income is better from a commercial agency than from some other method of selling.

However, there is a further problem. Fish marketing is one of those areas where the actions of individuals create external costs. One of the great dangers throughout the world of single-company fish selling is that hygiene standards will slide in the vicinity of the sales, to the detriment of the industry generally. The external cost arises because less fish is sold, or fish which is sold attracts a lower price, when the general hygiene standards of a market are poor. Regulation, by FPO or similar, government or municipality then becomes essential. The view taken in this text is that the future is not in this direction, but more towards regulation by FPOs.

Hygiene regulations

It is important that, from the outset, auctions take place under proper health and hygiene conditions. This is the direction in which the EU has moved and fish markets are having to modernise. Also, fish buyers increasingly want to purchase fish under hygiene and temperature-controlled conditions so that they can meet the conditions laid down by their own customers.

In developing countries the improvement of the conditions under which fish is sold is a real challenge, but must be done if the sector is going to be integrated into the modern economy. Finding the money to build and, perhaps even more difficult, to maintain to a high standard fish selling facilities is a problem in collective action (see Chapter 5). It will never be enough for individual users to keep the facilities clean because there will always be free riders. Official regulation in some form is almost certainly needed.

The legal entity established to operate the auction

The company set up to run the auction may be:

- A Fish Producers' Organisation
- A fish selling agency set up by an FPO
- A company owned jointly by an FPO and other parties
- A company owned by the port operator subject to the agreement of the FPO
- Some other commercial company under licence from the FPO
- A private company or individual which makes auctioning its or his business.

In all circumstances, however, it is desirable that the organisation should be properly established as a limited liability company, with a board of directors, having a legal right to operate as a fish selling agency. Hereafter, whatever entity or entities are set up, or apply, to function as a fish selling agency or agencies, are referred to as the Fish Auction Company or FAC. The issues that need to be considered are equally applicable to electronic or remote auctions (see Chapter 13).

Fish auction facilities

The facilities within which fish auctions take place may be owned, leased or rented by the entity established to run the auction. Often they will be owned by the harbour owners and used by the FAC. It is important that the facilities meet all the necessary standards and are regularly cleaned. Assuming that the FAC operates under licence from the FPO, the FPO must lay down very clear guidelines on what needs to be done and how it should be paid for.

Authority to sell fish

The FPO may license a number of different auctioneers or agencies to sell fish provided they adhere to the rules laid down. Alternatively, the FPO may decide that one agent is enough. It is a significant decision for the fishermen but will depend on the customs and practices of the port.

Internal organisation of the auction company

The company required to run an auction company is likely to consist of a small number of staff. However, ultimately the board of directors will have to take responsibility for the operation of the company. This point is worth emphasis (see Chapter 6) – directors have legal responsibilities for the orderly management of their companies.

Time of commencement of auction

The time of commencement of the auction is a matter for negotiation between the auction company and the fish buyers. The usual time in many fish auctions in the UK is 7 AM or 7.30 AM, but much depends on when the fish buyers need to start their processing staff working and the time they need to get the fish on the road to their customers.

Qualification of buyers

It is assumed here that the FAC will wish, in principle, to accept all buyers as bidders. However, the FAC may wish to consider the pre-qualification of buyers by means of an application form, and it may wish to require a bond or guarantee from buyers. This will imply that if ever a buyer fails to meet his financial obligations to the fish sellers the guarantee or bond can be exercised and at least some money recovered. The view taken here is that it should be open to the FAC to require some form of guarantee, but it is not essential. One problem is that providing bonds or guarantees through the banking system is not costless so it will be resisted by buyers. Also, the payments due for fish can easily be much greater than the value of the guarantee, so it serves little useful purpose. At the same time, the FAC may have sufficient knowledge of the buyers so that it does not need the guarantee.

Determination of lots

The auctioneer employed by the FAC, or working under licence from the FPO, sells fish by lot. A lot is defined as one or more boxes of fish of the same species, size and freshness. Boxes are generally of a standard size in most fishing ports.

Bidding procedure

The auctioneer invites price bids expressed in whatever is the standard unit of sale – normally a kilogramme but possibly another unit depending on custom and practice – for the fish in the lot for sale. Buyers may then increase their bids until there are no further offers. The 'one or all' principle is that the highest bidder is then allowed to choose how much of the lot he wishes to buy. The remainder of the lot is then offered for sale in the same way until all the fish in the lot has been sold.

  The 'one or all' sales principle is standard practice across Europe. The auction described is an upward price auction, and the buyer then chooses how much fish he wants at that price. This is known as the 'English' system. The 'Dutch' system works on a descending price, but is essentially the same. The buyers intervene at the price of their choice and then select how much fish they want to buy.

Buyer's responsibility for removal of fish

When fish is purchased in accordance with the agreed auction procedure the buyer is responsible for removing the said fish from the auction.

Name, address and proof of identity of buyers

The auctioneer needs to know who the buyer is. Normally this is achieved by local custom, but it may be necessary for the fish seller to take down details.

Payment terms

Payment terms are a major issue. Some fish auctions offer credit whereas in other cases payment has to be made very soon after the purchase, say no later than midday of the day of the sale. The payment made must, of course, include all taxes, levies and any other supplementary payments.

Credit terms

The FPO may decide that some formal system for managing credit is required. In Grimsby the 7-day rule is interpreted as meaning that settlings are all made on the Tuesday of the week following the transaction, so it can mean that payment may not be received until several days after the fish has changed hands. It is advisable in this case to secure a bank guarantee which can be called upon if the fish buyer defaults.

Buyers' responsibility for damaged fish

The buyer has to be responsible from the time of collection for loss or damage to lots purchased. The FAC cannot be responsible for any loss or damage of any kind whether caused by negligence or otherwise once a lot leaves its control.

Fish not taken away

Fish purchased but not taken away can be a problem, so the FAC needs to take the necessary precautions. It is suggested here that if fish duly purchased is not taken away the FAC should take steps to do one of the following:

- Proceed against the buyer for breach of contract
- Rescind the sale of that or any other lots sold to the defaulting buyer at the same or any other auction
- Resell the fish by auction
- Resell the fish by private sale
- Dispose of the fish in some other way.

The buyer still needs to be responsible for any resulting deficiency between the bid price and the subsequent receipts from the sale of the fish.

Ownership of fish by sellers

The seller must not put fish up for auction sale unless he is the true owner of the fish and is legally able to transfer title to it to buyers.

Commission

The FAC is entitled to charge an agreed commission for selling fish on behalf of the seller. The agreed commission will probably be in the order of 3% of the value of sales. However, it may be possible to operate on less.

Product quality

The FAC under licence from the FPO should ensure that all fish is correctly described in respect of species, weight, size and grade. However, buyers should satisfy themselves before the auction as to the condition of the fish on the market.

Failure to sell

When a lot fails to sell the FAC should notify the seller. The seller may then make arrangements to re-offer the lot for sale or to remove the fish from the market and pay any expenses incurred by the FAC.

Fish boxes of the FAC

The management of fish boxes is a major problem for the FAC. It needs to operate within a clear set of principles. The kind of arrangement that may work is that boxes supplied by the FAC should be returned by buyers to the company by 2 PM on the same day as the auction. The FAC needs to have some set of rules such as: boxes should be returned washed and in a good state of repair; and buyers of fish should meet the cost of repair or replacement of all damaged or lost boxes taken away from the auction.

Payments to sellers and sales notes

The FAC, having sold the fish, then needs to pay the fishermen. Again, an agreed principle is necessary. The suggestion here is that payment should be no later than

5 PM on the day following the sale of the seller's fish except when the sale is on a Friday, when payment may be made by 5 PM on the following Monday. Sellers should receive a Sales Note from the FAC, on which fish sales by quantity and value together with all deductions are listed and calculated, at the same time as payment for the fish is made. An important incentive to sellers to participate in the auction is prompt payment. If the FAC is offering credit to buyers and paying sellers immediately, it must have some working capital to cover the difference.

Fish sales other than by auction

The FAC should be entitled to sell fish other than by the defined auction procedure if the company judges that the returns to fishermen are thereby improved. However, it is important that transparency be maintained so a mechanism should be found to keep FPO members informed about what is going on, otherwise some of them may feel unfairly treated and become disillusioned with the system. The FPO is the appropriate entity for circulating information of this kind.

Information

The FAC should be responsible for maintaining a market information system. The suggestion here is that the FAC should ensure that the volumes of fish, by species, size and grade, expected to be available on the market are written on a blackboard or similar surface in a prominent position in the auction hall before the auction sale commences. The FAC should prepare a summary of the volumes of fish, by species, size and grade, at the end of each working day, and make this information freely available to buyers and sellers. The FAC should prepare a rolling forecast of expected supplies of fish and make this information freely available to buyers and sellers. The FAC should prepare monthly summaries of the volume and value of fish sales, by species, size and grade, and make this freely available to buyers and sellers. The FAC should prepare a daily list of all offers to buy fish outside the auction. The list should include the potential buyer's name, the species he wishes to buy, the quantity of fish requested and the price offered. The list should then be passed to the FPO.

Fish farming and fish auctions

Fish farmers are entitled to put their production through the auction system. Agricultural farmers use auctions to sell livestock, meat and crops so there is no reason on the face of it why fish farmers should not experiment with this method of sale. It would be important to notify buyers in advance that some farmed fish was going to be offered for sale to generate the interest required and then if it failed to attract a reserve price it could be withdrawn. The same considerations apply in this case as for other producers. *Prima facie*, opening buying to more competition might improve the price received for some farmed production, but if the architecture of farmers includes special relationships (i.e. repeated games – see Chapter 5), they might prefer to continue with other selling methods.

Summary

Establishment of Fish Producers' Organisation

FPOs, or similar catcher-controlled entities, are recommended as the main instrument for laying down fish selling rules on behalf of the industry. FPOs are voluntary associations of fishing vessel owners and operators formed in order to enter joint arrangements for the production and sale of their fish.

There are four regulated fish-selling options: (1) individual vessel sales; (2) contract sales of the FPO as a whole; (3) sale by auction; (4) private contracts. When dealing with the more homogeneous bulk products, such as herring or mackerel, it may suit the fish producers to sell under a contract. The location of auctions can be decided by agreement between producers and buyers. Hygiene regulations must be enforced.

The legal entity established to operate the auction

The company set up to run the auction may be:

- An FPO
- A fish selling agency set up by an FPO
- A company owned jointly by an FPO and other parties
- A company owned by the port operator subject to the agreement of the FPO
- Some other commercial company under licence from the FPO
- A private company or individual which makes auctioning its or his business.

Qualification of buyers

Pre-qualification of buyers may be desirable as a sifting process and the entity employed to auction fish may require the lodging of a guarantee.

Determination of lots

The term 'lot' is used to define one or more boxes of fish, of a similar size and freshness, and of the same species.

Bidding procedure

The auctioneer invites price bids expressed in whatever is the standard unit of sale

– normally a kilogramme but it may be another unit – for the fish in the lot for sale. Buyers may then increase their bids until there are no further offers. The 'one or all' principle is that the highest bidder is then allowed to choose how much of the lot he wishes to buy. The remainder of the lot is then offered for sale in the same way until all the fish in the lot has been sold. Under the English system, bidding goes upwards. The Dutch system works on a descending price. The buyer intervenes at the price of his choice and then selects his fish.

Payment terms

Payment terms are a major issue. Some fish auctions offer credit whereas in other cases payment has to be made very soon after the purchase, say no later than midday of the day of the sale. The payment made must, of course, include all taxes, levies and any other supplementary payments.

Commission

The auctioneer is entitled to charge an agreed commission for selling fish on behalf of the seller. The agreed commission will probably be in the order of 3% of the value of sales. However, it may be possible to operate on less, in which case a reduction is in order.

Fish boxes and information

The management of fish boxes is a major problem for the FAC, so it needs to consider very carefully how the system is going to be managed. The distribution of information on supplies and prices is another area which requires careful consideration.

Chapter 13

Establishment and Operation of Fishing Ports and Harbours

The establishment of satisfactory fishing ports and harbours makes a fundamental contribution to the development of a thriving private sector fish industry. In terms of the analytical framework discussed in Chapters 2 to 5, two broad points can be made. Harbours are a strategic asset for port owners and operators. Port owners need to consider what architecture is required if they are to maximise their returns. They can exploit the facilities for short-term gain – effectively asset stripping. Or they can take a longer-term view and use the assets for fishing or other commercial purposes. Or the third possibility is that the fishing harbour should be used to achieve a wider economic benefit for industry and the wider community. What happens depends very much on the strategic choices of the owner.

Second, fishing harbours are an important part of the environment of the fishing industry. They are the location where industry people engage in building up their architecture and reputation and where they learn about innovations. So harbour owners and operators who want to preside over a prospering fishing industry need to remember that the fishing port or harbour plays a crucial role in the creation of the commercial and social environment for a successful and prosperous fish industry.

Selected UK harbours: an overview

The first part of this chapter includes summary descriptions of the way different ports in the UK, Denmark and Poland operate. It will be evident that there are a variety of approaches in place. The chapter concludes with some suggestions on the preferred features of fishing ports and harbours for successful private sector development. Readers are invited to bear in mind the strategic implications of harbour developments, noted in the introduction to this chapter.

Fraserburgh

Special features

(1) The Harbour Commissioners responsible for the operation of the port have recently invested in new auction halls to conform to EU standards. This

development is matched by many similar developments throughout the EU – the drive to raise standards to meet the quality standards required by EU Directive 91/493/EEC.

(2) With the support of the fishermen, the Harbour Commissioners are developing deep-water facilities for the landing and processing of pelagic fish, in bulk. This is in recognition that Fraserburgh is one of the main EU landing ports for herring and mackerel.

Port management

The port of Fraserburgh is governed by a Harbour Commission, designated as the Harbour Authority by a special Act of Parliament. The first Harbour Authority for Fraserburgh was established by Parliament in 1815. Strictly speaking, the ultimate 'owner' of Fraserburgh is the state, but for all practical purposes responsibility for the proper management of the port rests with the Harbour Commission, with powers to invest, borrow and raise charges for the use of the port facilities. The Harbour Commissioners are the 'board of directors' of Fraserburgh Harbour. They are formally appointed by the Secretary of State for Scotland. In practice, they are suitable representatives of the local fishing industry and allied trades, which is significant in securing a cooperative response from all sectors of the industry and ensures that the Commissioners are focused on the industry's needs. The institutional setting encourages interested parties to play a cooperative game rather than the defection strategy of the Prisoner's Dilemma (see Chapter 5).

The most influential person in the management of the port is the Harbour Master. The structure of harbour management at this port has meant that it has carried out steady improvements to attract fishing vessels and to widen its scope for handling commercial vessels. The new fish markets, completed in 1987 and 1989, are just a relatively small part of the total effort by the port authorities to improve the commercial activity in the port. For example, a project currently in progress is to deepen the harbour entrance and one of the inner harbours so that large pelagic fishing vessels and commercial (non-fishing) vessels can enter.

The fish auction

The total auction market area is about $2000\,\mathrm{m}^2$. This is divided into two. One half is an overspill area. The main auction area has chilling equipment which reduces the temperature of the auction hall and enables fish to be landed at any time of the day or night. The auction itself is only for white fish (non-pelagic species) and takes place at 7.30 AM from Monday to Saturday. Prawns are also landed at Fraserburgh, but these and the pelagic fish are not sold through the auction. They are sold under direct contracts to buyers.

As in other UK ports fish is auctioned, or sold under contract, by fish selling companies operating on behalf of the fishermen. These take a commission (varying, but in the order of 3% of the gross sales value) for selling fish for the fishermen. The fish selling companies (known as agents) employ staff with auctioning skills. Box management is a potentially major problem (see Chapter 12) but is dealt with in an orderly manner in Fraserburgh. Boxes for fishermen to take to sea and land their fish are leased to fishermen by two specialist companies. After the sale of fish, boxes are returned by fish merchants to the agents of the fishermen who then put them in special box pounds for washing.

Grimsby

Special features

(1)   A new management company, Grimsby Fish Dock Enterprises Ltd, has leased the fish dock from the private sector harbour owners, Associated British Ports.
(2)   Grimsby has new fish market facilities.
(3)   Clear ultimate responsibility for the conduct of fish sales goes to the Grimsby Fish Producers' Organisation Ltd, delegated in practice to the fish selling agents in cooperation with Grimsby Fish Dock Enterprises Ltd.

Port management

Grimsby is of wider interest because a newly formed company, Grimsby Fish Dock Enterprises Ltd, has been established to lease the Grimsby fish dock from the owning company, Associated British Ports (ABP). ABP used to be The British Transport Docks Board and until 1983 was part of the public sector. It was one of the earlier privatisations and has been a story of almost unmitigated commercial success since then, if success is measured in terms of the profitability of the business – its strategic assets (its commercial harbours) have been operated almost exclusively for the benefit of ABP shareholders. However, any visitor to Grimsby fish docks would, until recently, have recognised the signs of generalised defection by many of those using the dock estate. Broken boxes, rubbish, decaying wharves, old-fashioned auction facilities and silting of the dock were all in evidence.

Having a dock owner from the private sector has probably been harmful to the development of its customers in Grimsby fish docks, the fishing industry. This is partly because ABP's profitability depends on being a successful rentier rather than operator of docks. It lets other businesses pay a rent and has cut back its own operating work. Some in the fishing industry have felt in the past that ABP was not a committed fishing port owner, in the way that one might say the Harbour Commissioners in Fraserburgh are. However, there is now recognition in the port that ABP has behaved generously and helpfully in the establishment of Grimsby Fish Dock Enterprises.

Managing the fishing industry of a port is more complex than commercial dock management because a great deal has to be organised. ABP has relinquished the management of the dock to a consortium, which includes representatives of the fishing industry. Grimsby Fish Dock Enterprises has leased the fish dock estate from ABP. The shareholders of Grimsby Fish Dock Enterprises include the Grimsby Fish Producers' Organisation, the Grimsby Fish Merchants Association, the local authority as well as a number of other investors from the private sector. Grimsby Fish Dock Enterprises has secured public sector finance (about £15 million) to enable it to renovate Grimsby Fish Market. The bulk of the cost was not the construction of the market itself. Rather, it was filling in part of the old fish dock to allow the construction of new facilities on reclaimed land.

The fish auction

The auctions in Grimsby (and in Hull) were originally set up by the trawler owners. Each vessel-owning company had its own team of fish salesmen who were responsible for selling the fish of the vessels of their employers. Following the decline of the company-owned fleet the concept of the vessel agency gained ground. Normally today, in Scotland as well as England, the vessel agent undertakes to sell fish from any vessel, whether he has an ownership interest in it or not, for a fee (normally about 3% of the sales value). He may also undertake various other vessel management tasks for the vessel owner. Nowadays, the rules governing the sale of fish are determined by the Grimsby Fish Producers' Organisation whereas in the past the rules were fixed by the Grimsby Fishing Vessel Owners' Association.

Plymouth

(Gillian Rodriguez, pers. comm. 1996.)

Special features

(1) The harbour (known as Sutton Harbour) is owned by Sutton Harbour Company (private sector).
(2) Plymouth Trawler Agents Ltd (the fish selling agency) is partially owned by the fishermen themselves (but not the Fish Producers' Organisation).
(3) Recently, a completely new fish market development was largely financed by grant aid.

Port management

In the UK fishing industry the Plymouth development is seen as a great success because the effect of the new market facilities has been to increase turnover from about £800 000 to over £2 million a year. Bunkering is seen by the harbour owners as a useful source of revenue.

Fish boxes are supplied by Sutton Harbour Company who make an *ad valorem* charge to the fishermen for box use. Boxes are cleaned for a small charge. The problem, as always, is recovering the boxes from fish buyers.

The fish auction

The fish auction in Plymouth is typical of the UK model, with the operation being wholly in the hands of the private sector, acting in close and friendly cooperation with the private sector owners of the harbour.

Peterhead

The port of Peterhead in Scotland operates in a similar way to Fraserburgh. It is run by the Peterhead Harbour Trust, set up by an Act of Parliament, which operates in a very similar way to the Harbour Commission of Fraserburgh. The Peterhead Harbour Trustees consist of representatives of local industry and commerce. The most impressive feature of Peterhead is the volume of fish traded; in 1994, for example, about 136 000 tonnes of white fish worth £84 million. Again, as in Fraserburgh, the port has benefited from a high degree of commitment by the Trustees.

Hull

The City of Hull is an industrial centre with a great variety of different industries represented. Food, including fish processing, is an important component. The Port of Hull is, like Grimsby, owned by ABP, and is mainly a commercial (non-fishing) port. The fishing section is only a small part of the total and landings of fish by local vessels are few.

A key feature of the Hull fish dock is its success in attracting wet fish from Iceland in containers for first sale on the Hull auction market. A second point worth noting is that, by tradition, the Hull auction is a 'Dutch' auction (descending price) unlike any other fish auction in the UK, which all operate the 'English' system of a rising price – bidding upwards for defined lots of fish.

Bridlington

Bridlington is a small fishing harbour about 30 miles from Hull. Like Fraserburgh, it is run by a Harbour Commission. However, it is noted here because the fishermen have seriously considered setting up auction in Bridlington, but have rejected the idea. They have decided to continue to use the Hull auction for selling fish which is not sold locally to restaurants, retailers and merchants.

Danish fish auctions

The Danish auction system is established under an Act of Parliament, so in that sense it is completely different from the liberal Anglo-Saxon model. The Auction Master is contracted by the government to undertake the task of running the auction. In this respect the system differs from the English system, under which auctions take place entirely voluntarily as a matter of commercial preference.

The system operates without an automated auction clock. Computers have been tried, and are used for preparing sales notes, but are not used in the auction process itself. A price incremental system is employed. In the case of the pelagic auction the minimum purchase for a buyer is ten boxes of fish. For the demersal auction the minimum purchase is one box of fish. So the system is like the English 'one or all' in the sense that the buyer, when he has selected a price, can buy as many units of the fish in question as he wishes.

Facilities are, in general, of a very high standard, extremely clean, without odour and well-maintained. These issues are very important in the long-term so that standards are maintained. The fee in Hirtshals for vessels is 2.46% of the gross sales for the use of the auction hall and this includes the payment to the auction master who receives 0.5% for his services. Harbours are owned by the public sector.

Keeping control of fish boxes is invariably a problem and always requires ingenuity by the management. The box system at Hirtshals is as follows:

- The Fishermen's Association owns the fish boxes and retains them in a box pool.
- The fishermen take their requirements from the box pool which keeps a record of this.
- The fishermen are charged 4 kroner per box for the rental.
- Fish buyers are obliged to return the boxes to the pool when they have been emptied.
- The cost of box rental is a deduction from the sales revenue, recorded on the sales note. The accounting is carried out by the auction master.

Herring is pumped into hoppers from a purse seiner, and then fed into 40-kg boxes ready for auction sales. This is of interest because it shows how some of the advantages of an automated system can be combined with traditional practice to meet the market requirement for boxed fish for human consumption. Herring is sold by auction or by contract in Hirtshals. The demersal fish auction in Hirtshals, takes place in a different location from the pelagic auction. Fish is graded and labelled in accordance with EU standards and is sold under hygienic conditions. The market information system is a notice board listing volumes of fish available to the port. There might be scope for improvement here bearing in mind that information reduces transactions costs.

A guarantee committing a bank to pay the debts of a fish buyer who fails to meet his commitments is common in the fishing industry. They generally take the form of a bond supplied by the bank to the Fish Producers' Organisation or fishing vessel owners' association. Guarantees are not in general used at Hirtshals because of sufficient trust between the auction master and the fish buyers.

Fishing ports in Poland

Location

There are 65 fishing ports of varying size and importance along the Polish coast (the Baltic Sea) (Lukomski, 1995; Palfreman *et al.*, 1996). The majority of these are simple anchorages. Around 20 are of significant size. The majority of ports are situated around river mouths (such as the mouths of the Odra, Swina, Dziwna, Parsenta, Wieprza, Słupia, Łeba and Wisła) which create naturally advantageous conditions for their construction. Some ports, such as Gdynia, Władysławowo and Swinoujskcie, became signficant fishing harbours for administrative convenience. The dominant category of fishing vessel in the Polish Baltic fleet is the 'cutter' – a small stern trawler 20–25 m in length.

Port administration

Legally, Polish fishing ports belong to the state and are administered under the Fisheries Department, until recently part of the Ministry of Transport and Maritime Economy, but now within the Ministry of Agriculture. Management and control is carried out in the three maritime offices of Szczecin, Słupsk and Gdynia. Following the Second World War, and the introduction of the command economy (whereby resources are allocated by administrative decision rather than via market signals), the ports were under the effective administration of vertically integrated fishing enterprises. The privatisation of the fishing fleet and the development of private sector fish wholesaling have brought this partially to an end. In practice, the degree to which management and control is devolved to the state-owned companies operating the individual ports varies. The probability is that the municipalities will take over responsibility for the harbours. The examples selected below illustrate that the harbours are essentially territories virtually owned by the vertically integrated state-owned company which is in place. It will also be fairly clear that the current arrangements are not as conducive to private sector development as they might be.

Władysławowo

The harbour territory of Władysławowo is owned and managed by the state-owned fishing and processing enterprise 'Szkuner'. Some premises are rented to other organisations, such as a fuel station, the Polish Register of Ships, the Navy and others. Eighty-eight cutters are based there, although there is some variation in exact numbers.

Szkuner has moved towards providing more commercial transparency in its general approach. It publishes the average prices on visible posters and in bulletins for the fishermen to read, so they know what they are likely to get if they land at Władysławowo. Szkuner is an example of one of the more forward-looking companies, which will probably be a success when it becomes privatised. However, the harbour estate is not open to competitive buying because of the control exercised by this one company.

Kołobrzeg

The fishing harbour of Kołobrzeg and a number of buildings, including the cold store, belong to the fishing and processing company 'Barka'. The ownership of the harbour territory (2089 ha) was accorded to Barka in 1994 by the Provincial Governor (Wojewoda). Previously the harbour was administered by the government.

There are 35 to 49 cutters based at the harbour. About 15 private companies are now operating in the harbour, supplying various services or processing fish. Like Szkuner, Barka has moved towards greater commercial transparency, by publishing prices to be offered for fish sold at Kołobrzeg. At present, it represents a significant commercial opportunity because reports from Poland indicate that Barka is in liquidation.

Ustka

The harbour of Ustka and its operations are managed by the state-owned enterprise 'Korab'. A similar number of cutters operate out of Ustka as in Kołobrzeg and sell to Korab. Korab rents out most of its property to 15 private processing and trading companies. Korab is the main user of the harbour territory and has initiated the procedure to obtain formal ownership.

Darłowo

The legal status of the fishing harbour and area at Darłowo is the same as Ustka. This port is managed by the state-owned fishing and processing company 'Kuter', which is a tenant of the Polish Exchequer. From 21 to 43 cutters operate from this harbour, who sell most of their fish to Kuter although some fish is also sold to outside enterprises.

Hel

Hel is managed by 'Koga', a state-owned enterprise. The total harbour territory is 180 296 m^2. Hel is effectively run by Koga, but other companies (cold storage and engineering) also operate there. The total number of cutters discharging fish at Hel ranges from 24 to 29.

Computerised auction systems

None of the ports and harbours noted above has installed a computerised system, although it is increasingly common in the Netherlands, France and Belgium, and a system has just been introduced at Milford Haven. So this account would be incomplete without some notes on them. Basically, the system has two components: (1) an auction 'clock' and (2) a computer to record the details of the sale of the fish from each vessel.

The clock with a figurative rotating arm, or rapidly changing prices up or down in some other electrical or electronic format, replaces the auctioneer who calls out the prices in response to bids. Buyers are connected remotely to the changing price and can stop it at the selected price. The details of each successful bid are manually entered into a computer which then calculates the sales return for the vessel, and within a few hours produces a sales note. Because the system has been developed in the Netherlands it is normally used for the 'Dutch' auction system – the arm on the clock falls until a buyer stops it at his selected price and then offers to buy some fish ('one or all' – anything from one box to the entire lot). However, it can also be used for the 'English' system, a rising price, because all that is being replaced is the auctioneer. Further advances in the technology now allow truly remote bidding from offices and homes of fish buyers, which has been adopted to a greater or lesser degree in ports in France, Belgium and the Netherlands.

UK fish merchants are still very tentative about using these clocks and trials have not so far led to any significant use. A computerised system which dispenses with a physical clock is now installed in Milford Haven (*Fishing News*, 1998). A projected computer screen replaces the clock and enables buyers in Zeebrugge and Milford Haven to buy in each other's markets.

There are several problems concerning the introduction of computerised systems, according to the UK fishing industry. The complexity of fish selling in Peterhead, with a number of sales going on at the same time, is one. The conservatism of the fish selling and fish merchanting communities in Hull and Grimsby is another. But also, because a great deal of buying and selling of fish already takes place over the Internet, and many fresh and frozen sales are now direct transactions between buyers and sellers, a number of people who are not technophobes still see little point in rigging up their markets with the modern technology when transactions seem to be increasingly avoiding the auction market altogether.

Impact of the Common Fisheries Policy of the European Union on operation of fish harbours

Concept of the fish producers' organisation

One of the most useful ideas to come out of the Common Fisheries Policy is the concept of the fish producers' organisation (FPO) (see Chapters 11 and 12). It is important to understand that these organisations are not simply trade associations or cooperatives. They are designed to meet a specific requirement, namely to strengthen the market power of the fish catching sub-sector. The argument is that of necessity this sub-sector is at a relative disadvantage compared to fish buyers. This is because the catchers are relatively dispersed when finding fish and are price takers when they put their fish on the market. Before FPOs, fishermen of the UK faced a continual legal problem if they wished to coordinate their landings. It was actually illegal, under competition law, for businesses to enter into restrictive trading agreements, such as an agreement to restrict supplies or fix minimum prices. There are now about 150 officially recognised FPOs within the EU.

Activities of fish producers' organisations

The scope for the activities of FPOs is wide ranging. The broad outlines are laid down in Council Regulation 3759/92. This Regulation states:

> *'producers' organization* means any recognized organization or association of organisations, established on producers' own initiative for the purpose of taking such measures as will ensure that fishing is carried out along rational lines and that conditions for the sale of their produce are improved.
>
> *Article 4*
> These measures ... *require* members:
> to dispose, through the organisation, of their total output of the product or products in respect of which they are members; the organization may decide that this requirement can be waived if products are disposed of in accordance with common rules....
>
> ... to apply ... rules ... with the particular aim of improving quality, adapting the volume of supply to market requirements and ensuring proper management of catch quotas authorised.
>
> ... to apply, where the Member State concerned has provided for the management of some or all of its national catch quota or quotas to be managed by producers' organizations, within the limits of the quantities allocated to the Member State from the total allowable catches for the stock or group of stocks in question, the measures necessary to ensure the proper management of the catch quotas authorized.'

Application of withdrawal prices

One of the main roles of FPOs in fishing harbours is to apply withdrawal prices. If the fish withdrawn is one of the official species (including herring) the FPO is obliged to pay compensation. It receives the support it needs to do this from intervention funds up to approximately three-quarters of the withdrawal price. So the FPO has to have its own funds as well as relying on official intervention funds. The scheme is also designed so that the compensation system works in emergency, but does not allow fishermen simply to fish for compensation. In other words, the scheme is designed to encourage fishermen to become masters of their own fate, but it does not provide unlimited subsidies for fish that the market does not want.

For illustrative purposes, the 1998 official withdrawal prices for cod are set out in Table 13.1.

Table 13.1   1998 Official (EU) withdrawal prices for cod (ecu/tonne).

Species	Size class	1998 Withdrawal prices			
		Gutted fish with head		Whole fish	
		Extra, A	B	Extra, A	B
Cod	1	1063	1004	768	590
(Gadus morhua)	2	1063	1004	768	590
	3	1004	827	590	472
	4	791	543	449	421
	5	551	319	331	171

Application of marketing rules

FPOs can, if they have the determination, take over responsibility for marketing. In the case of the Scottish Fishermen's Organisation the FPO is acting as a salesman on behalf of its members and it has also invested in fish processing capacity for dealing with bulk landings of pelagic fish. The Grimsby FPO applies marketing rules. The Grimsby FPO rules deal with matters such as responsibility for grading, the agencies that are licensed to sell fish on behalf of members, the requirement to auction fish, units of sale, and, most important, the requirement upon members to supply the FPO with logsheet and sales information.

Quota management

When FPOs were first established in 1973 it was envisaged that the fisheries management carried out by FPOs would be the adjustment of quantities caught to meet market requirements. However, it soon became clear that the legal framework also gave FPOs an opportunity to manage quotas in a way acceptable to the membership. Today FPOs play a significant role deciding the rules under which national quotas are managed.

EU hygiene regulations

Hygiene regulations are laid down in the Council Directive of 22 July 1991 which specifies 'the health conditions for production and placing on the market of fishery products' (91/493/EEC). This is quite a complicated regulation. Chapter II of the Annex to the Directive is important because it specifies requirements during and after landing. In summary only, the regulation requires the following (this list is not definitive and those requiring a guide to the conditions to be met when planning fish markets should refer to the Council Directive or their own hygiene authorities):

- The creation of a list of approved establishments authorised to handle and package fish products.
- Allocation of health and hygiene inspection to a competent authority.
- That fish or fish products originating in third countries should come from establishments approved by rigorous checking of quality, hygiene standards and documentation.
- That unloading and landing equipment must be constructed of material that is easy to clean and disinfect and must be kept in a good state of repair and cleanliness.
- During landing contamination of fishery products must be avoided. Operations have to proceed rapidly, temperature must be appropriate, with ice used if necessary, and equipment should not damage the edible parts of products.
- Parts of auction or wholesale markets where fishery products are displayed for sale must:
  - be covered and have walls that are easy to clean;
  - have waterproof flooring which is easy to wash and disinfect in such a way as to facilitate the drainage of water and have a hygienic waste water disposal system;
  - be equipped with sanitary facilities with an appropriate number of wash basins and flush lavatories;
  - be well-lit;
  - exclude exhaust-producing vehicles and undesirable animals;
  - be cleaned regularly, at least after each sale. Boxes and crates must be cleaned and rinsed inside and outside with drinking water or clean sea-water and, where necessary, disinfected;
  - have signs prohibiting smoking, spitting, eating and drinking;
  - be closeable and be kept closed when the competent authority considers it necessary;
  - have adequate water supplies;

- have special watertight receptacles made of corrosion-resistant materials for fishery products that are unfit for human consumption;
- have a lockable inspection room.

• That if fish is to be stored before transport cool rooms must be available.
• That suitable clean work clothes and head gear be worn.

Market information systems

Information available in fishing ports and harbours has been touched on in this chapter. In many ports the harbour authorities attend to the issue of supplying key commercial data whilst in others the private sector provides the service. It is an issue which those concerned with fishing harbour development should consider because of the impact of information on transactions costs. Information might be made available in accordance with the suggested simple format of Fig. 13.1.

Vessel	Mon	Tues	Wed	Thurs	Fri	Next week
Mary Anne		100 kg herring 100 kg sprat				
Queen of Sheba	500 kg herring 100 kg sprat 100 kg cod 100 kg others					
Etc.						Nil (storm forecast)
Totals						

Fig. 13.1   A simple system for forecasting supplies.

Daily summary of prices

The daily summary of prices might take the form of a simple note on paper or writing on a blackboard.

Monthly summaries

A possible format for preparing monthly summaries is given in Fig. 13.2.

Summary

At Fraserburgh and Peterhead the combination of committed technical and administrative skills supplied by the Harbour Commissioners and representation

	Month			Year to date		
	Volume (tonnes)	Value (total)	Av. value (price/tonne)	Volume (tonnes)	Value (total)	Av. value (price/tonne)
Cod EA 12						
Cod B 12						
Cod B 3						
Cod EA 4						
Cod B 4						
Cod EA 5						
Total cod						
Herring 1						
Herring 2						
Total hearring						
Total sprat						
Etc. – to suit requirements						
Others						
Total						

Fig. 13.2   Monthly summary of landings and prices.

of the industry by relevant bodies in the port is precisely the mix that leads to good harbour management. The tendency of a minority of fishermen to defect – for example, to fail to observe auction rules or to use their vessels in an irresponsible way in the harbour area – is overcome by the committed and authoritative stance of the Harbour Commissioners. They have also made great efforts, often very successfully, to attract grant assistance for the improvement of both the fishing and commercial sides of harbour business.

Fish does not have to be sold by auction. Pelagic fish and prawns are both sold in Fraserburgh, but by direct contracts between the buyer and sellers (the fishermen) and not through the auction. This is because there are relatively few buyers and both buyers and sellers want to move the fish as quickly as possible. These products are sold subject to producer organisation rules. In the case of herring and mackerel one of the main buyers is the fish producers' organisation itself (the Scottish Fishermen's Organisation).

The story of the Humber ports shows that it is important for the vessel owners and operators and the port owners to reach a *modus vivendi*. Whatever arrangements are reached for running a fishing harbour, it is essential that the various interests in the industry work together to resolve the various problems which are bound to arise in a busy industrial context.

Grimsby shows that the fish producers (the catching sub-sector) can regulate the market without necessarily owning the fish selling companies or the harbour estate.

In Plymouth, fish selling is undertaken by a private sector business, partially owned by some of the fishermen. The fish producers' organisation is not significantly involved. It shows that many different patterns of ownership and responsibility can emerge. Harbour ownership committed to the success of the fishing activities is the crucial ingredient.

The examples of Polish ports show that the re-organisation of the industry into a transparent, competitive sector has some way to go. The market power of the current harbour controllers is still likely to stifle rather than encourage development. The harbours need to be opened up to outside competitive forces.

Danish as well as British experience shows that fish merchants are still very tentative about using computerised auction systems. There seem to be no dramatic cost savings or they would have been adopted more widely by now.

The Danish experience demonstrates that the liberalisation of the EU has influenced Danish fish-selling procedures. In the EU fishermen are under no obligation to sell through the auction system, whereas, before the EU system was enforced, Danish fishermen were obliged to use it. A guarantee committing a bank to pay the debts of a fish buyer who fails to meet his commitments is common in the fishing industry. Denmark shows that when conditions of trust prevail these are not always necessary.

Fishing ports need an information system. The suggestion here is that three sets of data should be collected, analysed and made widely available:

(1)   A forecast of landings (by species) for the following week, rolled forward day by day as information improves.
(2)   A daily summary of prices of fish, published at the end of each auction.
(3)   Monthly summaries of the landings and prices by FPO members (*not* individual vessels, which might be commercially sensitive data).

Modern fishing ports and harbours need to be planned and managed with hygiene standards in mind.

The concept of the fish producers' organisation has been particularly helpful in the creation and sustaining of orderly marketing of fish, whether it takes place in a fish market or over the Internet.

Chapter 14
Planning and Management of Fisheries Sector Projects

Projects and change

In business a project is a concentration of resources used over a limited time scale to bring about change (see Chapters 3 and 7). Planned projects are part of the business plan. The business plan may be a project in itself, or it may include projects that the company intends to undertake. Projects should generate returns – at least to cover the capital and operating costs involved, but preferably adding something more to the value of the business, the economic added value referred to in Chapters 1–4. To achieve this, the business should be able to identify a distinctive capability (architecture, reputation, innovation and strategic assets) which it expects to turn into a competitive advantage in the marketplace.

Projects are a concentration of resources used up over a limited time horizon, with a fixed beginning and end. A project is intended to achieve something (the project objectives) within a given time scale. This book does not deal with project planning and management techniques for the fisheries sector in depth, but, as so much of management requires the projectisation of development, it does provide an outline of the main points.

The project cycle

All projects are governed by the project cycle. Chapter 7 outlines the stages of business inception which quite clearly are the same as for project start-up, but sometimes with different labels. International development agencies tend to use the World Bank methodology (Baum, 1982; Baum and Tolbert, 1985). Under this scheme the cycle is broken down into six stages:

(1)   Project identification
(2)   Project preparation
(3)   Project appraisal
(4)   Project implementation
(5)   Project monitoring
(6)   Project evaluation.

Project *identification* is the stage in the cycle when project ideas are spotted and screened. This is equivalent to stage (1) of the business inception scheme in Chapter 7. The detailed work of project planning is *preparation*. *Appraisal* is the final review of the plans, often undertaken by a different team of consultants in the case of projects funded by development agencies. This is equivalent to stage (2) of the scheme. The long and expensive part of the cycle takes place at *implementation* [stage (3) of the scheme – lift off]. *Monitoring* is the normal project management activity to check on its progress towards its objectives. The backward look at the end has come to be known as *evaluation* and is an attempt to see what can be learned from the experience for future projects. Both of these are important for proper project management so fall into the stage (4) of business inception.

Commercial project management specialists often employ different terminology, but the cycle essentially follows the same route (Turner, 1993, 1995). Projects are conceived and then planned in detail. Implementation is the time-consuming and most costly part of the cycle. Commercial project managers have to have control routines, so they need to monitor the implementation. At the end, there is the point known as close-out when the project comes to an end.

Project management

This book does not include a detailed exposition of project management techniques. Readers are referred to Turner (1993, 1995) for an excellent and practical exposition. Logframes are not covered by Turner but can be reviewed in a number of development manuals, for example, the Commission of the European Communities' *Manual of Project Cycle Management* (1993).

However, it is worth noting that there is a body of knowledge which has been borrowed from the project management sectors such as defence and construction, and used increasingly by mainstream businesses to manage projects. The sector where the project management approach has been particularly useful is Information Technology where very clear project methodologies are essential to business planning if total confusion is to be avoided.

A selection of useful project management techniques and concepts is given below.

The project triangle

In the project triangle the project takes place within three constraints, denoted by the corners of a triangle. These are time, cost and specification.

The logical project framework

The logical project framework is a tool used by aid and development organisations to force project planners into the discipline of specifying and clarifying project objectives, summarising the inputs required to achieve those objectives and showing how the project will be monitored.

The milestone plan

The milestone plan is a model of the project organised into stages, perhaps between 10 and 20. Each of these represents a subsidiary objective, otherwise know as a deliverable, a work package, or a product, which the project must achieve if it is to move on to its final aim.

The bar chart

Bar charts (or Gantt charts, after the name of their inventor) are bars placed on a graph to show the sequence and length of the various work packages required to achieve each milestone.

The responsibility chart

The responsibility chart sets out responsibilities with respect to the project, including 'single-point authority' for overall project management as well as for the various intermediate milestones in the project. Some form of dual reporting system is almost inevitable for private sector projects, unless they are large. The project team is seconded to the project, but they are highly likely to have other line responsibilities which they need to keep going whilst they are working on the project, hence the phrase 'dual reporting' because project staff report both to their line manager and to the project manager. However, a project manager who is responsible for the project is essential.

Statement of monitoring responsibilities

To ensure that project monitoring is undertaken for control purposes, someone or an institution has to be responsible for it.

Defined and well-understood 'change management' procedures

Change management is an important part of project management because things very rarely go exactly according to the original plan. A formal agreed procedure is needed for changes to the original plan which will inevitably crop up.

Fish businesses

The private sector in the fish industry consists of a wide range of businesses, ranging from single-person part-time catching or merchanting enterprises to large multinationals. There are also cooperatives and non-governmental organisations (see below). In many capture fisheries there is a genuine conflict of interest between the private sector and the economy as a whole. This is because of the common property character of the resource. On the one hand, private companies are interested in profit. The state, on the other hand, must consider the interest of the economy as a whole, which is often in conflict with private businesses (Clark and Munro, 1975, 1980; Cunningham *et al.*, 1980; Clark, 1990). This is why some degree of regulation of the private sector in fisheries is essential.

Projects and cooperatives with Non-Governmental Organisations (NGOs)

Cooperatives are usually part of the private sector. They have been seen as a means by which relatively poor fishermen or farmers can assert market power, improve their standing in society or provide important economic services at lower cost. However, the record of fisheries cooperatives throughout the world is not good. This means that great care must be taken when considering projects involving them.

The phrase 'non-governmental organisation' or 'NGO' is now used to refer mainly to institutions set up to implement projects funded by charitable organisations. (The Appendix to Chapter 4 describes a Bangladesh example of an NGO in some detail.) Because they are usually set up to assist the very poor they can be a useful, ready-made means of reaching this section of the community. Moreover, NGOs often have programmes of their own projects. In fisheries, NGO projects can often go wrong because of the absence of expertise. This is regrettable as well as sometimes being avoidable. NGO projects generate external costs in a similar way to other projects, and new developments can operate to the detriment of existing NGOs.

Framework of laws and regulations

The framework of laws and regulations which sustain and govern the fishery, maritime and commercial operations is an important background to fisheries and aquaculture projects. It sets out the parameters within which the project must function and the legal restraints on its activities. The legal and regulatory areas which should be checked off before embarking on a project related to fisheries or aquaculture are discussed below.

The waters for the fishery

The legal status of the waters from which the fish supplies relevant to the project are expected to come should be established. In the case of sea or ocean waters it is important to know if the fish is expected to be caught in open access areas outside national jurisdiction, from within the exclusive economic zone, from within the territorial sea, behind base lines or within some protected zone, or some combination of these. It is also important to establish whether waters can be closed or are likely to be closed for military purposes, such as target practice, or other commercial purposes, such as oil or gas drilling or gravel dredging.

Similarly, if the project relates to inland waters their legal status should be clarified. This applies to aquaculture projects as well as to projects for rivers and lakes where laws can often be unclear.

Species

International and national regulations may apply to the species with which the project is concerned. These should be identified and it should also be established whether any changes are impending. The laws and regulations dealing with prohibitions on species, fish size regulations and their implications for the project, gear restrictions, vessel restrictions, seasonal restrictions and catch quotas are all relevant. These regulations can have an impact on fish processing and distribution projects as well as on catching sector projects, and a regulation on one species can have an impact on the catching or prices of another. Thus, whatever a project is intended to do, an assessment of the legal background is likely to be important.

Operation of vessels

For catching sector projects, in marine or inland waters, it is important to establish the rules governing safety on the water, the safety equipment vessels must carry, shipping regulations to the extent that they impinge, nationality rules, insurance rules, navigation rules, pilotage arrangements, and regulations governing port arrivals and departures.

Handling of fish

There will be laws or regulations governing fish or food handling. There may be grading requirements, rules on prices, quality control systems and obligations to land by selected persons such as registered dock workers to selected institutions, such as state trading organisations.

Employment of persons ashore and afloat

Regulations concerning the qualifications, training, nationality and registration of seamen will exist. There will also be rules dealing with shore-based employment which are relevant. Working hours, the employment of women, national insurance contributions, personal taxation, training, the status of trade unions and demarcation of work may be subject to regulation. See also Chapter 6 for some comments about employment. We emphasise, however, that employment law and regulation is a minefield. Business and project developers need to exercise caution in ensuring that they follow the appropriate regulations.

Operation of factories

Some of the country's factory legislation may be relevant to the operation of plant at sea or on shore.

Fish farms

Fish farming operates within its own regulatory framework in many countries. Governments are concerned about matters such as pollution and the control of diseases in fish.

Environmental regulations

All countries have laws and regulations that are designed to protect the environment which must be checked before any commercial decisions are made.

Access to official research

Information relevant to commercial fishery operations may be governed by secrecy legislation. It is important to establish to what information management will have access.

Company and other relevant institutional laws

The operation of company law or other institutional law may be important as background to the project itself as a legal entity with its own legal status. For example, if fishing operations are bound by company law, accounting and registration requirements, there will be a flow of information which can be used as input data for the planning and management process.

Tax laws

All relevant taxes, exemptions and rebates must be established. This is important in the description of the legal environment as well as for the financial appraisal.

Fisheries projects and the environment

The environment is the natural world. It includes many features which are taken for granted, such as the air and oceans. It also includes features of which people are increasingly aware, such as beautiful scenery and forests. It includes natural resources which are not immediately visible, such as underground mineral reserves. Fisheries are also part of the given environment, which is why fisheries studies are also environmental studies and fisheries economics is a branch of environmental economics. The environment is humanity's natural capital stock. Like any other capital *stock*, the natural world, or environment, provides a flow of benefits for us to use. The economics of the environment is concerned with explaining why, so often, society does not make the best use of environmental resources and with managing the environmental stock such that the flow of benefits is at its best possible level for mankind.

When projects are formulated, an assessment of the extent to which they deplete or enhance the capacity of the environment to continue to generate a flow of benefits will often be a requirement. In many cases this study of the environment, the 'environmental impact assessment', will become a major, and possibly decisive, part of the project analysis and will be required by the authorities before a private sector project can go ahead. Even without this statutory requirement many businesses undertake an assessment of the environmental impact of their plans, perhaps on ethical grounds, but it is also increasingly perceived as contributing to public relations.

Environmental impact assessment

Environmental impact assessment (EIA) is an *assessment* of the environmental impacts of a selected development. A detailed description of environmental impact analysis in aquaculture appears in Midlen and Redding (1998). There are numerous general works on EIA, for example Dixon *et al.* (1986). It is essentially a scientific undertaking, but it is sometimes possible to add on an economic analysis. EIA does the following:

- Identifies
- Quantifies
- Predicts
- Evaluates.

EIA is not simple and its sheer complexity helps to explain the popularity of the 'precautionary principle'. In many cases environmental impacts are impossible to trace though any reasonable observer is aware that environmental damage is taking place. One way of dealing with the problem is to take a strategic view and to assume that environmental damage will take place unless it can be proved otherwise. FAO has produced a Code of Conduct for Responsible Fisheries (FAO, 1995) which embodies the application of the precautionary principle to fisheries, and then followed the Code with a series of Technical Guidelines for each sub-sector.

An example

To illustrate the complexity of EIA a brief summary of what might be involved in an analysis of an agro-industrial development on an estuary is presented.

(1)   The first issue is the prediction of the scale of the industrial or agricultural activities themselves. In many industrial projects it is possible to predict the scale of the activity although the outcome may be different from the prediction. For agricultural activities, forecasting the input and output decisions of farmers is also extremely difficult. In principle, however, this step is part of the project analysis and should be possible.

(2)   Step 2 is more problematic. The emissions of pollutants or other environmental damage must be estimated. This is a complex technical task which can only be carried out by experienced people. Moreover, if the private sector is involved, the companies concerned have an incentive to underestimate the emission of pollutants.

(3)   Step 3 is harder still. The degree of concentration and the spatial distribution of the pollutants in the environment must be estimated. In the case of an estuary, water is moving through it all the time, dispersing, suspending or dissolving the pollutants.

(4)   At step 4 the physical consequences for people and animals (including fish) must be considered. It may not be possible to judge whether the pollutants in the expected concentrations are likely to accumulate in the biological populations who use the water body.

(5)   In the case of fisheries, the EIA would look for an estimate of the impact of the accumulation of pollutants on the physical productivity of the resource. It may also be relevant to assess the effect on consumer valuations. In the latter case the consequences of pollution can be highly significant. For example, the sedentary species of an estuary, such as the molluscs, can easily be rendered inedible through pollution. Even if this is not so, the subjective valuations by consumers of fish that might be contaminated can be drastically influenced.

(6)   Finally, the economist would attempt to assess the economic cost of the pollution. This would have two components:
  (a)   the cost of the fall in productivity. The expected loss of output would be one element of the cost of the pollution
  (b)   the fall in consumer valuations. This is a cost because it represents a fall in the value of output. Various techniques for arriving at valuations are now available to economists.

Social aspects of fisheries projects

Social analysis is also sometimes a pre-condition for private sector fisheries projects. This is basically a check that the outcome of a project is not detrimental to equity. Techniques are described in Baum and Tolbert (1985) and Asian Development Bank (1991).

Estimation of costs

One of the main lessons to emerge from the evaluation of fisheries projects is that both capital and recurrent costs are frequently underestimated at the project planning stage. This may be due to simple practical omissions: outdated price lists are used or quotations that are unlikely to be applicable at the time of the procurement of inputs are obtained. In addition, the costs of transport to remote sites are often difficult to obtain and are then underestimated. Similarly, insurance costs may turn out to be higher than planned. Sometimes the technical services needed to get a project into the state where it is operational are underestimated.

In the case of operating costs planners often fail to appreciate all the costs that have to be incurred for normal commercial operations. Sometimes the omissions concern the commercial skills needed to make a project a success: the costs of accountancy, legal, administrative, company secretarial and marketing skills are not fully taken into account during project planning. More generally, the scope for things simply failing to run according to plan is not appreciated.

Degree of precision in project costing

How precise does project costing have to be? There must be sufficient information provided to ensure that each project activity and capital element is described in adequate detail for it to be clear how the project is designed to work. In general, therefore, the project planner should seek to provide adequate detail and be as precise as possible, but inevitably there will be fuzzy areas. In the case of a fish collection (from artisanal fishermen), processing and marketing project, for example, the planner might be able to specify in considerable detail the equipment required. This might include ice-making and crushing equipment at the landing points, insulated vehicles to carry the fish from the landing points to the processing plant, the size and equipment of the processing plant and perhaps details of the refrigerated vessels required to ship the finished product. In this case the business would be able to cost with precision all the capital cost items, the raw material inputs, labour and energy requirements and product outputs over the life of the project. In contrast, the designers of a project to sell a new type of beach-landing craft to a number of fishing communities may not be able to predict with any certainty the adoption rate. It will depend critically on the confidence of the fishermen in the new craft.

The guidance in Table 14.1 may be helpful.

Table 14.1 Degree of precision in project costings.

Estimate	Accuracy	Source
Ball-park	± 25%	Initial judgement based on experience. Likely to be used at project identification
Comparative	± 15%	Based on more research, but still preliminary. May be used for cost estimate by contractor
Feasibility	± 10%	The result of project preparation
Definitive	± 2%	After design work finished. Sometimes not fully available until project is underway

Cost concepts

There is frequently confusion concerning what should be regarded as a project cost and the distinction between costs and finance. Whilst not attempting detailed definitions, this section discusses some of the more important concepts. These concepts often overlap.

Capital investment

In the process of investment, capital is combined with the other factors of production, such as land and labour, to produce a stream of outputs for the economy to use. Real capital is the goods and services collected together in the process of investment. The word 'capital' is used to define the goods and services put together for a particular project. Notice that the word 'services' is used alongside 'goods' as part of capital. Investment requires services, such as the transportation of the equipment to the factory, or the services of engineers to set it up. Some investment involves very little physical capital, and is almost entirely service, such as the investment in human skills when people attend a management course. The money value is the common denominator which stands for the real goods and services. When we undertake the analysis of projects we measure capital as this common denominator – the amount of money it costs.

Variable costs and fixed costs, direct and indirect costs

A behavioural definition of costs is to divide them into fixed and variable costs. Management accountants sometimes call 'variable costs' 'marginal costs' (Chadwick, 1991). Economists take a somewhat different view, as explained below.

Variable costs

What actually goes into variable costs depends on the time period over which the decision is to be taken. So, strictly speaking, for a project decision, all the costs, including capital costs, associated with the project must be taken into account, so all the costs are variable for the purposes of the business decision in question. However, the concept is used mainly in the context of business decision making in the short to medium term rather than project decisions. A fishing vessel owner might ask, therefore, what the variable (or incremental or marginal) costs are of putting a vessel to sea on any given day. His decision depends on whether the expected return meets the variable costs, although in the longer term he would look for more of a contribution to overheads.

Fixed costs

Fixed costs are those costs that have to be paid out irrespective of the level of activity of the enterprise. Again, fixed costs depend on behaviour and what is included in them depends on the time period of the management decision. So, if a period of a year was under review, fixed costs might include managerial salaries and other management costs, interest and loan repayments and rental payments.

Direct costs

Direct costs are costs that can be identified as forming part of the product or service, such as wages and raw materials used in the production process.

Indirect costs

Indirect costs are also known as overheads. They are expenses which do not form part of the product. They might include management salaries, the wages of cleaners and materials used for general maintenance.

Incremental (marginal) costs and sunk costs

Incremental costs

It is important to remember when preparing a project for funding the distinction between those costs that would be incurred in the absence of the project and those that arise as a direct result of the project. The latter are sometimes also described as 'incremental costs' or 'marginal costs' because they are *added to* the costs of the business because of the project, and are therefore an *increment* to the company's total costs.

For example, in a project designed to reduce a factory's post-harvest spoilage, workers may be retrained and further workers may be recruited. The project costs include the costs of moving the extension workers from other activities and the costs of employing new workers, but they do not include the departmental overheads that would have been incurred anyway in the absence of the project. The project costs are the incremental costs associated with the project. This is what

economists mean by incremental or marginal costs, which can obviously sometimes include what a management accountant might want to categorise as a fixed
cost.

Sunk costs

Sunk costs are 'sunk'. If certain costs have already been incurred before the project
is established, the general principle is that they should be ignored. For a business,
however, costs incurred in the past (such as an investment in a building), although
they may be sunk, may still generate fixed costs in the management accounts, as
depreciation and other inescapable outlays, such as interest and loan payments.
They would still be costs that the well-run company would wish to cover.

Suppose, for example, there are alternative sites for a fisheries harbour. One is
inconveniently located and will result in higher costs for the fishing industry. The
second is better located and results in a lower cost. However, several million
dollars have already been *sunk* in the first site and 'it seems a pity to see it all go to
waste'. Poker players would call opting for the first site on these grounds
'throwing good money after bad' and would not do it; neither should the fisheries
planner. The future has to be decided in its own right, not with reference to past
decisions, however good or bad.

Opportunity costs

Opportunity costs are costs of inputs in their next best alternative use. In most
cases this is the same as the market prices of inputs. However, for some inputs,
such as buildings or investments, the opportunity cost is what the enterprise (or
project) does with the asset if not used for the purpose envisaged. The opportunity
cost of a fishing vessel could therefore be its value if it were sold abroad. If
marginal or incremental costs are correctly estimated, including all the costs
implied by a decision at the margin, they are identical to opportunity costs. In the
case of a marginal decision, if much needs to be done to achieve it (such as setting
up a new company, buying land, building a factory, employing and training a
workforce), the marginal/opportunity costs are high. If, however, the work
involved is low (such as adding to the stock in a cold store), the marginal/
opportunity cost is low. The point illustrates how it is not always satisfactory to
identify variable costs with marginal costs.

Depreciation

Just as there are two senses in which the word 'capital' is used, so there are two
corresponding senses in which 'depreciation' is applied. For economists, depreciation is the reduction in the value of an asset (such as a machine) over time
resulting from wear and tear through use or obsolescence. For accountants,
depreciation is an accounting concept designed to inject a measure of the loss in

the value of the asset ('capital consumption') into the measurement of annual corporate performance.

Evidently there is a very close correspondence between the two concepts. In any event, depreciation is not included in the cash flow projections used in project analysis because it does not stand for any movement of real resources used up by the project. The value of the capital is already included at the outset, and if some or all of that capital is replaced at some later stage in the project, then the cash paid out to buy the new capital will also be included to stand for that movement of real resources. Capital expenditure is already counted and it would be double-counting to add depreciation as an additional cost.

Although depreciation is not included in the cash flow analysis, it appears in the profit and loss account so that readers of the accounts can see, year by year, how much capital the company is consuming. Tax liabilities are usually reduced by deducting depreciation costs from profits. Through the profit and loss account, depreciation influences the tax liability (if any) of the enterprise and is, therefore, of interest to the sponsors or owners of the project.

Interest and loan repayments

Interest and loan repayments are also not included in the basic cash flow statement, but for a different reason. One of the principal reasons for deriving the cash flow is to determine the rate of interest and loan repayments that the project is capable of supporting. The discounting calculations described in Chapter 15 are a method of comparing the project with the costs of capital. To include interest and loan repayments in the cash flow as well as discounting would be a form of double counting.

An 'after financing' cash flow may be prepared in addition to the basic investment appraisal cash flow. This does include interest and loan repayments as deductions from the cash flow. This 'after financing' statement is to assess the capacity of the project to meet its financial commitments and to place a value on the return from the project to its owners ('the return to equity').

As in the case of depreciation, the costs of interest payments appear as a cost ('above the line') in the profit and loss account.

Main categories of project cost

For the purposes of financial analysis, project costs are those goods and services for which there is a financial outlay. It is convenient to consider them in two groupings, *capital costs* and *operating costs*. Capital costs are incurred when the initial capital assets of the project are being put into place and, again, when they are replaced. Operating costs, sometimes called recurrent costs, are the expenditure on goods and services which is needed for the operation and maintenance of the project.

Fish industry projects

A list of headings of items that would be part of a typical simple investment project in catching, processing and aquaculture is given in the Appendix to this chapter. A detailed itemised list would be included in the appropriate place in the feasibility study.

Guidelines for cost–benefit analysis

A computer model which simulates the project should be developed. This is, in part, the exercise of amassing all the available technical information and seeing how it fits together. The planner should find out how the project inputs are expected to produce the project outputs. This exercise requires technical advice, or it may even be done by a technical expert rather than an economist or an accountant. It is, however, more than a technical exercise because, to function, the project elements must be coordinated, legal requirements must be fulfilled, the labour force must be adequate for the job and offices must function efficiently. So, in addition to the purely technical information, the planner must think managerially. He must ask what additional inputs of physical capital, training or management skills are essential to the project.

The technical characteristics of the project may be clear, but it is always worth enquiring whether there are technical alternatives to the proposed plan. Can some, or all, of the project outputs be achieved in some other technically feasible way? For example, is it possible to run the project less intensively, with fewer inputs, and lower levels of output? Are there alternatives to the type of investment proposed for fishing vessels? In general, the planner should also be building up alternative options to the project.

The next step is to assign values to the project inputs. This is a question of establishing, as accurately as possible, the costs of the inputs at the project location for each year that the project is expected to run. It may entail some ingenuity and effort to obtain all the necessary data. Moreover, some care must be taken to ensure that *all* the costs are included. For example, for a warehouse, the transport costs, including a labour element, must be added to the operating costs. If the point of landing fish is at some distance from the fish market an allowance for additional transport costs must be added.

This chapter advises planners to ignore price increases which are the result of general inflation. However, relative price increases or decreases should be included. So, if the planner finds out that the cost of one or more of the inputs is expected to rise (or fall) over the lifetime of the project, he should include this intelligence in his analysis of operating costs.

For a business, the benefits from an investment project are the additions to the worth of the business following from it. The businessman hopes that his risk and expense will be rewarded with a stream of net cash surpluses stretching into the future. The problem is to forecast what sales can be expected.

Each column in Table 14.2 incorporates different illustrative assumptions. Columns 1, 3 and 4 demonstrate variants on the case of a fillet yield of 45%. In columns 1 and 2 the price of the fillets is $5000 per tonne, in column 3, $4500 per tonne and column 4, $4550 per tonne. With a fillet yield of 45%, the price of $4550 is the *switching value*, that is, the price at which the net benefit turns from positive to negative (column 4). The incremental cost of $50 000 might be the cost of employing additional fish process workers as well as additional packing and marketing costs. This example shows how, in principle, any project to enhance the value of production should be approached. The benefits in the form of the extra value the fish can be sold for should be offset against the extra costs of enhancing its value. In this example only the first case (column 1) indicates any significant benefits from undertaking the extra work of filleting the fish.

Table 14.2   Cost–benefit analysis of gains from filleting fish.

	Predicted annual results			
	1	2	3	4
Whole grouper supplies (t)	1000	1000	1000	1000
Filleted weight (t)	450	400	450	450
Current sales of whole grouper ('000$)	2000	2000	2000	2000
Sales price of filleted grouper ($/tonne)	5000	5000	4500	4550
Predicted sales of filleted grouper ('000$)	2250	2000	2025	2050
Incremental revenue ('000$)	250	—	25	50
Incremental cost ('000$)	50	50	50	50
Net benefit ('000$)	200	−50	−25	0

*See text for explanation of columns 1–4.

Real life cases may be more complicated. For example, for the column 1 case let us abandon the assumption that no investment cost is required. It may be necessary to purchase blast freezers and cold stores, packaging equipment and materials and the workforce may need training. To illustrate let us assume that freezing equipment costing $187 000 is purchased. Installation costs are $3000. In addition, a further $10 000 must be spent on training the workforce in filleting skills and in using equipment. Thus, the total capital cost is $200 000. It is assumed that the capital cost is incurred in 1 year and the project has a life of 10 years. The project is evidently attractive. The investment cost of $200 000 would be recovered in 1 year and the internal rate of return (see Chapter 15) works out at 100%.

Benefits of cost reduction

A project that reduces cost generates benefits to the extent of the net cost reduction. For example, investment in a harbour facility may reduce vessel operating costs by providing lower cost services, safer berthing and perhaps scope for a more rapid turn-around in port. The estimation of benefits is a question of calculating the net saving accruing to the vessels (and any other harbour users) and then, for the harbour owner, judging how much of the saving can be recovered in increased landing charges.

Other project benefits

In fisheries projects, as in most other 'production' sectors, benefits normally arise from increases in the value of production (higher prices and/or volume) or cost savings within the project boundary. There are a number of other headings under which benefits are sometimes attributed to projects by people trying to sell the concept (such as employment, food, multiplier effects, technological and educational benefits). All these external benefits are, strictly speaking irrelevant to the businessman unless they are internalised into the project calculation. Even so, it is never good practice to attribute to projects benefits for which there is no plausible case or economic rationale.

Summary

Projects and change

For business a project is a means of bringing about change. Projects are a concentration of resources used up over a limited time horizon, with a fixed beginning and end. A project is intended to achieve something (the project objectives) within a given time scale.

The project cycle

All projects are governed by the project cycle. Under this scheme the cycle is broken down into six stages: (1) Project identification (2) Project preparation (3) Project appraisal (4) Project implementation (5) Project monitoring (6) Project evaluation. Commercial project management specialists often employ different terminology, but the cycle essentially follows the same route.

Project management

Some useful tools and techniques used in project management include:

- The project triangle
- The logical project framework

- The milestone plan
- The bar chart
- The responsibility chart
- A statement of monitoring responsibilities
- Defined and well-understood 'change management' procedures.

Framework of laws and regulations

A checklist of areas to consider:

- The waters for the fishery
- Species
- Operation of vessels
- Handling of fish
- Employment of persons ashore and afloat.
- Operation of factories
- Fish farms
- Environmental regulations
- Access to official research
- Company and other relevant institutional laws
- Tax laws.

Fisheries projects and the environment

The study of the environment, the 'environmental impact assessment', is likely to become a major, and possibly decisive, part of project analysis.

Environmental impact assessment

Environmental impact assessment concerns the following in relation to the environment: the identification, quantification, prediction and evaluation of impacts.

Social aspects of fisheries projects

Social analysis of projects may also be required.

Estimation of costs

One of the main lessons to emerge from the evaluation of fisheries projects is that both capital and recurrent costs are frequently underestimated at the project planning stage.

Degree of precision in project costing

The degree of precision in project costing depends on when in the project cycle the estimate is made and the requirements of the project sponsors.

Cost concepts

- Capital investment
- Variable costs and fixed costs, direct and indirect costs
- Incremental (marginal) costs and sunk costs
- Opportunity costs
- Depreciation
- Interest and loan repayments.

Guidelines for cost–benefit analysis

A computer model which simulates the project should be developed. Furthermore, the technical characteristics of the project may be clear, but it is always worth enquiring whether there are technical alternatives to the proposed plan. The next step is to assign values to the project inputs. This is a question of establishing, as accurately as possible, the costs of the inputs at the project location for each year that the project is expected to run.

Benefits of cost reduction

A project that reduces cost generates benefits to the extent of the net cost reduction.

Other project benefits

In fisheries projects, as in most other 'production' sectors, benefits normally arise from increases in the value of production (higher prices and/or volume) or cost savings within the project boundary.

APPENDIX

Table A 14.1   Checklist of capital items for a fishing vessel investment project.

Item	Comments
Vessels	Number, registered length, overall length, gross registered tonnage, fishing methods for which designated, construction materials, engine size and type
Fishing gear	All details of nets or other catching equipment and hauling gear
Ancillary equipment	Fish-finding equipment, radar, safety gear, refrigeration, ice-making and other equipment
Vessel delivery	Time and location
Preparation for use	Requirements for skilled personnel, such as maintenance engineers, to prepare vessels for commercial use
Boxes	Stock of fish boxes included
Packaging materials	Packaging materials, such as cartons and polybags as well as machinery for a packaging operation on board or on shore
Spares	Engine and other spares
Vehicles	Refrigerated and other vehicles plus all relevant details
Shipping and insurance	All transportation and insurance details
Land	Land, invariably required for gear storage, workshops, offices and sometimes housing
Buildings	Housing, offices, storage, workshops
Formation costs	The legal entity to run the project must be established
Training	If the project depends on trained local management some investment costs ('investment in human capital') may be necessary
Project design costs	
Professional fees	Estimate of professional input, such as accountancy, architecture, chartered engineers
Working capital	A real cost properly attributable to the project, but may be necessary to impute or estimate
Other	

Table A 14.2 Checklist of capital items for an industrial fish processing project.

Item	Comments
Land	Location is critical.
Buildings	Evidence that issues of product flow, work patterns, delivery systems are considered. Configuration of cool rooms, cold stores, freezers, ice makers and any other equipment important
Fish processing equipment	Can consist of blast and plate freezers, cold stores, cool rooms, block, crushed and flake ice makers, filleting equipment, canning machinery, band-saws and many other items. Investment depends on a very clear understanding of the nature of expected output and its market
Vehicles	Refrigerated and/or others
Boxes	Plastic, polystyrene, wooden: choice to be justified
Packaging materials	Cardboard cartons, expanded polystyrene boxes, wooden or plastic boxes, polybags, custom printed materials, etc.
Spare parts	Spares for machinery, chemicals for ice makers, supply of handling equipment
Micro-computers	Essential for accounts, pay-roll, sales records and analysis
Formation costs	The legal entity to run the project must be established
Pre-project training	
Design and management	
Professional fees	
Shipping and insurance	
Working capital	
Other	

Table A 14.3 Checklist of capital items for an aquaculture project.

Item	Comments
Civil works	
Land	Location is critical
Earthworks	
Drains	
Roads	
Buildings	Evidence that the issues of product flow, work patterns, delivery systems are considered
Machinery/equipment	
Hatchery	
Pumps	
Harvesting gear	
Packaging materials	
Laboratory equipment	
Boxes	Plastic, polystyrene, wooden: choice to be justified
Spare parts	
Packaging	
Computers	Essential for accounts, pay-roll, sales records and analysis
Transport	
Vehicles	Refrigerated and/or others
Services	
Formation costs	The legal entity to run the project must be established
Pre-project training	
Design and management	
Professional fees	
Shipping and insurance	
Working capital	
Other	

Chapter 15

Financial Analysis of Fisheries Projects

This chapter is an introduction to the financial analysis of fisheries projects. We can say that if a project earns more than the opportunity costs incurred in establishing it, including the opportunity cost of capital, then it contributes to EAV. The experience of this author suggests that financial analysis is done to validate the conceptual analysis. A project has to work as a coherent concept. The initiator needs to be able to demonstrate the project's distinctive capabilities, how these translate into competitive advantage and then EAV.

There are many books on accounting and projects which will provide a much more comprehensive survey of the accounting techniques needed for thorough analysis (for example: Gittinger, 1982; Higson, 1986; Chadwick, 1991; Curtis, 1994). Limitations of space prevent the inclusion of much detail here.

Valuing projects

Financial and economic analysis

A distinction should be drawn between financial and economic analysis of projects. Financial analysis treats the project as a business proposition and assesses whether or not it adds to the net worth of the business. If the net present value (NPV) of the project is expected to be positive then from the point of view of the business the project should be undertaken because it will add to the value of the company. For a whole variety of reasons, especially in fisheries, financial analysis does not indicate whether or not the project makes a net contribution to the wealth, or value, of the economy as a whole. For that, economic analysis is required, which is not a topic dealt with in this book. It is enough to state here that one of the main reasons for the divergence in fisheries is that the individual businessman does not normally take into account the impact of his investment on fisheries resources in his financial calculations. The state has to do that in its own calculations and secure the stocks through fisheries management measures.

The arithmetic of present value

Financial decisions have effects which are spread through time. The effects of an investment decision are a stream of positive or negative cash flows. The money the business spends on the vessels or processing plant or ponds, the operating costs, such as repairs, maintenance and replacement and wages, and the money it will receive for the output from the investment are the cash flows. Money has a time value known as the rate of interest, that is, the rate at which it can be borrowed or lent. This rate is used to *discount*, or reduce the value to the business of, the future stream of net cash flows.

The explanation for this is that if $100 is invested today at a rate of interest of 10%, then 1 year later it will be worth $100 \times (1 + 0.1) = $110. If $110 is invested for a further year at the same rate, it will be worth $110 \times (1 + 0.1) = $121. In other words, after 2 years the money is worth $100 \times (1 + 0.1) \times (1 + 0.1)$, that is, $100 \times (1.1)^2$.

The process of reinvesting capital and interest in the way described above is called compounding. The outcome of the process ($121 in the above example) is called the future value of the investment. In symbols:

FV = future value
PV = present value
r   = rate of interest (expressed as a decimal)
n   = number of periods for which the sum is invested
$FV = PV(1 + r)^n$.

Each year an investment generates a future value. In the example, the future value at the end of the first year is $110. Its present value is $110/(1 + 0.1) = $100. Similarly, the FV at the end of the second year is $121. The present value of this is $121/(1 + 0.1)(1 + 0.1)$ or $121/(1.1)^2$. The future values are, therefore, discounted at an annual rate of 10% to arrive at the present value. In symbols, for any given year:

$$PV \frac{FV}{(1 + r)^n}$$

The present value of a stream of future cash flows is simply the sum of the present values calculated from future values by this formula. The use of the word 'net' implies simply that costs have been deducted from benefits before the calculation starts. Suppose a project of $1000 is expected to generate a net cash flow of $300 per year for 4 years. As an alternative, the money could be deposited in the bank at 10% per year. Therefore, to judge whether it is worth investing the money in the project or whether it would be better to leave it in the bank the cash flow is discounted at 10%. Table 15.1 illustrates the position.

Under the assumptions, the investor would be better off leaving his money in the bank because the NPV is negative. However, the NPV is only a little less than zero, so let us assume that the investor finds a way of reducing annual operating costs in such a way that the net cash flow from years 1 to 4 increases to $320 per year. The discounted cash flow may then be recalculated, using the same discount

Table 15.1 Discounted cash flow: example (1) (in $).

Year	Cash flow	Discount factor	Discounted cash flow
0	(1000)	$1/(1.1)^0 = 1$	(1000)
1	300	$1/(1.1)^1 = 0.9091$	273
2	300	$1/(1.1)^2 = 0.8264$	247
3	300	$1/(1.1)^3 = 0.7513$	225
4	300	$1/(1.1)^4 = 0.6830$	205
Present value of cash flow stream			(50)

factors. Table 15.2 shows that the small change in the net cash flow now produces a positive NPV of $14. It now looks as though our investor would choose to go ahead with the project rather than leaving his money in the bank.

There are some points to note:

(1) It is normal practice to assume that the year of the major investment in the project is year zero. This is a matter of convention.

(2) Note also that 1.1 to the power of zero [$(1.1)^0$] does not, on the face of it, make any sense, but in mathematics it is assumed to equal 1, and it can be seen that this does give the correct answer.

(3) The discount factors listed above are calculated to demonstrate the arithmetic of discounting. In practice, this is rarely necessary as there are books of tables with these listed (for example, Lawson and Windle, 1970) for use with hand-held calculators. Also, with the advent of micro-computers, NPVs and similar statistics can be calculated very rapidly.

(4) Note that *cash flow* is used as the description of the future events to be discounted. Time is now a critical factor in decision making and it is important to identify if possible when costs and benefits arise. For decision-making purposes costs are incurred when cash is committed and benefits are realised when cash flows in.

(5) The sequence of cash flows in and out of the business is also the sequence of costs and benefits incurred by or accruing to it. This means that some accounting conventions, such as allowances for increases in property values and depreciation, are not included in cash flows. In the first of these cases, only when the cash on the sale of a property is realised does the value of it

Table 15.2 Discounted cash flow: example (2) (in $).

Year	Cash flow	Discount factors	Discounted cash flow
0	(1000)	1	(1000)
1	320	0.9091	291
2	320	0.8264	264
3	320	0.7513	240
4	320	0.6830	219
Net present value			14

enter the cash flow. In the second case, only when a new machine is purchased to replace an old machine does cash leave the business and is, therefore, included in the cash flow when it happens. However, the changes in asset values are included in profit and loss statements and balance sheets.

(6) In project analysis the custom is to ignore the effects of inflation and to work in real values. This implies that inflation affects the costs and benefits equally. It also means that the project analyst must find out the *real* (as opposed to the money) discount rate.

A decision rule for projects is, therefore, that if the NPV is positive (greater than zero) then the project should go ahead.

Internal rate of return

The internal rate of return (IRR) is the rate at which net benefits are discounted to give an NPV of exactly zero. In the example in Table 15.1 the IRR is equal to 7.7% and in the Table 15.2 example the IRR is 10.7%. The decision rule would be that if the project is expected to generate an IRR above the test rate of discount it should be accepted.

The IRR is a popular decision rule for two reasons. First, it is readily understandable as a rate of return. It is, therefore, easily saleable to administrators and politicians who may not be familiar with the NPV idea. Second, decision makers or planners may not have to hand an agreed rate at which the stream of net benefits can be discounted. Nevertheless, they can be fairly sure that a high IRR will be above it, whatever it is. There are, however, problems with it.

It is unsuitable for the evaluation of mutually exclusive alternative projects. Consider two mutually exclusive projects, one large, the other small. In the fish context these could be for large and small fishing vessels. The small project might be relatively low cost and show a high IRR compared to the large project. However, at the same time the large project might, in the end, have a significantly larger NPV and make a bigger contribution to the company's profits. So if the company has to do one or the other, as it might if it is constrained by licensing rules, then the small vessel choice would be the wrong one using an IRR criterion. This point might offer a simple explanation why businesses invest in large vessels although the IRR on small vessels often seems to be high – the large vessel makes a larger contribution to the company's profits. Also, in some special cases when the cash flow fluctuates between positive and negative over the life of the project, there may be no single IRR.

These issues can be pursued in any text on business finance (for example, Higson, 1986, Chapter 6). In practice, it is unlikely to be the technicalities of the method of calculating a decision rule that produce the wrong answers, but poor project identification and preparation.

Cost–benefit ratio

The cost–benefit ratio is the discounted stream of benefits divided by the discounted stream of costs. If the ratio is greater than one this implies that the gross benefits exceed the gross costs, or that the NPV is greater than zero. The decision rule would be to accept projects with a positive cost–benefit ratio. As with the IRR, in the case of mutually exclusive projects, its application could theoretically give the wrong answer. A small but profitable project with a high cost–benefit ratio might be chosen in preference to a larger project that in the end generated more value.

Payback criterion

The payback rule is that projects are ranked according to the payback period. Those whose positive cash flow meets the cost of the investment in the shortest time are the projects to be accepted. It works by taking just one element, the length of time to the recovery of the initial investment, making it paramount and ignoring all others. It is, clearly, a simple rule of thumb, and is undoubtedly popular with businessmen who want to get their money back as quickly as possible, and would see all subsequent profits as pure gain. Payback is not academically respectable and is unsuitable for public decision making largely because it discriminates against longer-term projects. Nevertheless, if it did not usually produce the decisions which businessmen want it would not have the appeal it does to even the most sophisticated of businesses.

Accounting rate of return

The accounting rate of return (ARR) is the accounting profit of the project divided by the book value of the assets employed in it. Again, this is popular with businessmen as it provides a simple indication of profit in percentage terms. It is used by businesses and one must assume, therefore, that, in general, it provides acceptable answers. There is a mathematical relationship between the IRR and ARR, and they are related concepts. The decision rule for a business based on the ARR is that projects showing a return above some predetermined rate would be accepted. This is very similar to the IRR rule.

General comment on decision rules

Given a bad set of projects even the most sophisticated decision rule will choose a bad project, but stick a pin into a set of good projects and one cannot fail to select a good project. If a business has resources to invest in either improving its decision techniques or improving its performance in generating ideas for new projects, it might do better to choose the latter.

Valuing projects under uncertainty

The literature on investment appraisal often draws a useful distinction between risk and uncertainty. The word 'risk' is used for those cases where the probabilities of the various possible outcomes (cost and benefits streams) are known, or at least there are good grounds for assuming that the probability distribution of outcomes follows a specific pattern. When much less is known about expected outcomes the situation is described as 'uncertain'. Whilst risk analysis is of some academic interest, and can be applied in advanced, highly developed financial markets, it is not, in general, applied to project planning and especially so in the case of fisheries where outcomes are bedevilled by extreme uncertainty. It is worth noting, too, that major industrial enterprises do not use risk analysis when they evaluate their capital investment proposals (Higson, 1986, Chapter 9).

Authorities usually recommend 'sensitivity analysis' as a means of handling the issue of uncertainty. This means simply trying out plausible alternative values for the variables in the appraisal and seeing what the effects are on the project as a whole. We see how sensitive the project is to changed assumptions. With modern micro-computers it is very easy to do. In fisheries projects the variations in variables such as catch rates or prices can be very substantial so the analyst must be bold in the options he tests.

A spin-off from sensitivity analysis is 'switching values'. Each variable in the financial model can be altered at will by the planner. The switching value is the value of any variable (price, catch rates, wages, rent, etc.) at which the project becomes unacceptable. That is, it is the value at which the NPV becomes negative or the IRR falls below the test rate of discount.

Planning of capital structures

The capital structure of the project is the ratios and amounts of each type of funding used to pay for it. The sponsors of the project are the people who own it. The equity, or shares, belong to them and the project is their ultimate responsibility. In addition to putting in some of their own resources (the equity), the sponsors may also opt to raise some of the capital they need by way of a loan. Interest will be payable on the loan and what remains after loan repayments and interest is known as the *return to equity*.

Equity can take different forms. If the government simply gives the assets to the institution selected to undertake the project, then that is the government's equity. If the project is run by a company, then the equity will be in the form of shares subscribed to the shareholders. Loans can also be raised in a variety of ways, such as from commercial banks, development banks or from private individuals.

Outline example of the financial analysis of a fisheries project

This section summarises the procedures of financial analysis by means of an imaginary worked example. The project is a vertically integrated fisheries development consisting of four trawlers targeting shrimp and other species together with some associated onshore facilities, and a possible increment of two trawlers. The detailed description of the project, its justification, a statement of objectives and the many other components of a typical project preparation report are omitted.

Currency units

The example is expressed in US dollars. The use of the most common hard currency comes easily to most fisheries specialists as much of the industry is international, so there is at least the advantage that the use of the dollar is very convenient as a widely understood yardstick of value. However, sometimes it will be more convenient to work in a local currency and there is no objection to this. It simply means that to arrive at the values needed to test for international competitiveness, the internationally traded goods and services must be converted back into local currency at an appropriate rate of exchange.

The tables are expressed in thousands of dollars. This is merely for convenience. We are dealing with variables about which there is a great deal of uncertainty so there is no need to introduce a spurious accuracy. However, the calculations for the tables were done in dollars and cents, not in thousands of dollars. The spreadsheet program was then instructed to divide the dollars by 1000 and to round the values.

The use of a micro-computer

The example was prepared on an IBM compatible micro-computer using a modern spreadsheet package program known as Microsoft Excel. It is not easy to compute the type of analysis that follows to the standards required by most investors without the assistance of a spreadsheet program and a micro-computer. Sensitivity analysis, that is, testing the results of various different assumptions concerning the variables, becomes a very simple matter. However, the author does not believe that the speed and efficiency in the manipulation of data are any substitute for hard conceptual and practical thinking about the project. Data can too easily be adjusted to make a bad project appear financially sound.

Cost and revenue assumptions

The project is assumed to have a life of 10 years although some of the capital components may have a longer physical life than this, and some a shorter. A possible increment to the project, part way through its life, is also analysed.

Investment schedule

Table 15.3 sets out the costs of the project investment together with the costs of the possible increment noted above. The capital investment listed in Table 15.3 constitutes the concentration of resources, that is, the beginning of the project. The project feasibility study will have explained, in detail, the content and justification for each of the items.

Table 15.3 Project capital costs (in '000 $).

Capital costs	Foreign	Local	Total	Increment
Jetty construction	1 000	500	1 500	
Buildings, etc.	2 000	1 000	3 000	
Refrigerations and electrics	500	0	500	
Processing equipment	150	50	200	
Mechanical handling equipment	200	0	200	
Pallets	0	100	100	
Packaging	250	0	250	
Hygiene and safety equipment	100	0	100	
Equipped vessels (4 + 2)	10 000	0	10 000	5 000
Vessel delivery	500	0	500	250
Engines and spares	1 000	0	1 000	500
Maintenance engineer	50	0	50	
Shipping and insurance	2 000	0	2 000	1 000
Working capital	3 000	0	3 000	
Pre-operating expenses	1 000	450	1 450	
Contingency	0	2 000	2 000	
Total	21 750	4 100	25 850	6 750

For example, the feasibility study will have explained why a jetty is required as part of the project. It will have given details of its scale, building materials, labour requirements and any other associated cost. Similarly, the item 'Buildings' will have been supported by a detailed explanation of the project's needs and a physical description of the buildings planned. The study will have included details of the refrigeration equipment to be purchased, the numbers of each item and the recommended product specifications. Note, however, that the finer details of the investment specification would probably have awaited the implementation phase of the project when procurement commenced.

Project investment costs are separated into local and foreign expenditure. This allows the project's owners to identify the direct foreign exchange commitment implied by the project and also provides a measure of the local commitment to it.

Working capital of $3 million has been included in the capital cost. This will pay for inputs, such as labour and fuel, which will combine with the fish catch to generate the outputs of frozen shrimp and other processed fish. If working capital must be injected into the project at the beginning, then, logically, in the final time period of the project outputs will be produced from invested working capital which have already been purchased in earlier time periods. A sum equivalent to the working capital is, therefore, added back into the net revenue stream in Year 9 (see Table 15.7).

Table 15.4 Project operating costs (in '000 $).

	Year 1	Year 2	Year 3	Years 4–9	Increment
Variable costs					
Labour[a]	30	40	50	50	15
Electricity[b]	30	35	40	40	20
Water[c]	6	8	10	10	4
Repairs and maintenance[d]	50	70	100	100	20
Fuel	1 000	1 000	1 000	1 000	500
Packaging material	100	200	300	400	200
Sales cost	10	10	10	10	5
Total variable costs	1 226	1 363	1 510	1 610	764
Fixed costs					
Salaries	665	665	665	665	283
Licences and royalties	10	10	10	10	5
Administration	10	10	10	10	0
Travel	100	100	100	100	0
Insurance	1 000	1 000	1 000	1 000	500
Project management	100	100	0	0	0
Housing maintenance	0	100	100	100	100
Legal and audit	5	5	5	5	5
Total fixed costs	1 890	1 990	1 890	1 890	893
Total costs	3 116	3 353	3 400	3 500	1 657

[a] Refer to Table 15.5(a).
[b] Refer to Table 15.5(b).
[c] Working at full capacity, annual consumption of onshore facilities (for ice and fish processing) will be 12 000 m^3. Annual consumption of the vessels will be 4000 m^3. Cost is 62.5 cents per m^3. It is assumed that in Years 1 and 2 annual consumption will be 60 and 80% of full capacity consumption, respectively.
[d] From Year 3, overhaul and hull repaint per vessel per year costs $20 000 per vessel plus $20 000 for onshore equipment plus other vessel repairs and maintenance.
The assumptions made to derive all other cost estimates should be explained to the reader.

Pre-operating expenses are also included in the investment schedule. If it is intended that expenses will be incurred before the project gets underway, then these are a part of the 'concentration of resources' which takes place at the outset.

The right-hand columns of Table 15.3 and 15.4 includes a possible increment to the project.

Operating costs

Table 15.4 sets out the operating costs for the project. Note that the table distinguishes between variable costs and fixed costs. Variable costs are those that vary with the scale of the operation and can readily be identified with a unit of production. In the long run *all* costs are variable, but in the short run (such as over a fishing season) some costs are fixed and inescapable.

Each of the items listed in Table 15.4 must be supported in the feasibility study by details of the assumptions on which it is based. Table 15.5(a) itemising labour costs and Table 15.5(b) giving details of electricity costs are examples of these. In some cases it may be possible to outline the assumptions upon which an item is based by a footnote to the table. Two examples of this practice are given. The remainder of the tables appears without supporting assumptions, although an adequately documented project document would supply them.

Table 15.5(a)   Annual labour cost at full capacity[a] (in '000 $).

	Number of men employed per year	Unit cost	Total cost	Increment Number of men employed per year	Unit cost	Total cost
Production	7	5	35	2	5	10
Vessel servicing and discharge	3	5	15	1	5	5
Total casual labour			50			15
Management	1	20	20	0	20	0
Engineers	2	18	36	0	18	0
Finance	2	15	32	0	15	0
Production	3	6	18	1	6	6
Administration	3	10	30	1	10	10
Vessels: skippers and engineers	8	15	120	4	15	60
Vessels: crew	32	12	384	15	12	192
Miscellaneous	5	5	25	3	5	15
Total non-casual labour			665			283
Total labour costs			715			298

[a] In years 1 and 2 casual labour inputs are assumed to be 60% and 80% of full capacity.

Labour costs

Table 15.5(a) sets out the project's annual labour costs plus supporting assumptions. The table is, of course, incomplete; the feasibility study would include a description of the project as an institution. The expected labour requirement for each year of its life would be described and analysed.

Some of the costs (salaries) of employing people are included under 'Fixed costs' (see Table 15.4) because they are inescapable over the time horizon of the project. The presumption is, therefore, that labour is employed under contractual arrangements, which implies that the numbers cannot be altered in the short run. If crews for vessels are employed on a casual basis it would be more appropriate to include the costs under variable costs.

Table 15.5(b) Electricity usage and cost.

	Year 1	Year 2	Year 3	Year 4
Ice plant (kWh/day)	250	250	250	250
Days	90	100	125	125
Total kWh	22 500	25 000	31 250	31 250
Cold store (kWh/day)	150	150	150	150
Days	280	295	300	300
Total kWh	42 000	44 250	45 000	45 000
Freezer (kWh/day)	300	300	300	300
Days	35	60	80	80
Total kWh	10 500	18 000	24 000	24 000
Grand (total kWh)	75 000	87 250	100 250	100 250
Price ($/kWh)	0.40	0.40	0.40	0.40
Total costs ($)	30 000	34 900	40 100	40 100

Revenue assumptions

Table 15.6 sets out the assumptions behind the revenue projections used for the project analysis. A forecast of revenue for fish production projections has two components: a forecast of prices and a forecast of catch rates. Both can go wrong. To arrive at satisfactory estimates of future prices an analysis of supply and demand for the commodity in question should be conducted. In this case the prices for which the project outputs may be sold are assumed to remain constant, despite increased supplies. This means that the size of the project output relative to the market as a whole is small. This may be a reasonable assumption for many fisheries projects, but it needs to be supported by a careful and analytical description of the relevant markets.

Catch rates (catch per unit of fishing effort) are notoriously difficult to predict. A prediction of high and profitable catch rates extending over time in an open access fishery is usually untenable. For profitable catch rates to persist some degree of effort limitation is a pre-condition of fisheries development. In the case depicted the presumption is that some form of effort limitation is in place to prevent excessive fishing effort from being attracted into the fishery.

Financial Analysis

The cash flow statement

The information presented above is sufficient to allow a simple projection of the cash flow to be prepared. The cash flow is what it says – the predicted flow of cash in exchange for goods and services flowing into and out of the project. Bearing in mind that we aim only for as much accuracy as is necessary to do the job, it is convenient to assume that the cash flow takes place on the last day of each year.

Table 15.6 Project revenue assumptions.

	Year 1	Year 2	Year 3	Years 4–9	Increment
Capacity (%)	60	75	90	100	
Export shrimp					
Tonnes pa	240	300	360	400	150
Price ($/t)	8 000	8 000	8 000	8 000	8 000
Revenue ($)	1 920 000	2 400 000	2 880 000	3 200 000	1 200 000
Export fish					
Tonnes pa	1 560	1 950	2 340	2 600	1 000
Price ($/t)	2 000	2 000	2 000	2 000	2 000
Revenue ($)	3 120 000	3 900 000	4 680 000	5 200 000	2 000 000
Local fish					
Tonnes pa	3 000	3 750	4 500	5 000	1 500
Price ($/t)	500	500	500	500	500
Revenue ($)	1 500 000	1 875 000	2 250 000	2 500 000	750 000
Ice sales					
Tonnes pa	600	750	900	1 000	0
Price ($/t)	50	50	50	50	50
Revenue ($)	30 000	37 500	45 000	50 000	0
Total revenue ($)	6 570 000	8 212 500	9 855 000	10 950 000	3 950 000

The example in Table 15.7 gives a net present value (NPV) of $2443 000 calculated at an annual discount rate of 12%.

If 12% is the opportunity cost of capital, the project will make a positive financial contribution to the financial position of its owners. Table 15.8 calculates the 'return to equity'. It demonstrates, therefore, what remains for the project's owners after they have met their financial commitments to the bank and to the government as well as the project's normal operating costs. The owner's financial commitments are shown in Tables 15.9 and 15.10. This information will be particularly important if private sector investors are involved. If it turns out that the return is insufficient to induce them to put money into the project, it may be possible to adjust the schedule of loan repayments to give them a higher return. The question of the distribution of the returns from the project is not addressed in this example.

Incremental costs and benefits

For the financial appraisal of any proposed investment, only the incremental costs and benefits associated with that investment are relevant. Overheads costs, which would have been incurred anyway in the absence of the project, are not relevant to the decision and are, therefore, excluded from the analysis.

If this example were under consideration by a corporation, the only costs and benefits relevant to the decision would be those directly associated with the

Table 15.7 Calculation of financial internal rate of return (in '000 $).

	Year 0	Year 1	Year 2	Year 3	Year 4	Year 5	Year 6	Year 7	Year 8	Year 9
Cost of investment	21 400								1 850[a]	
Working capital	3 000									−3 000[b]
Other capitalised investment	1 450									
Operating costs										
Variable		1 226	1 363	1 510	1 610	1 610	1 610	1 610	1 610	1 610
Fixed		1 890	1 990	1 890	1 890	1 890	1 890	1 890	1 890	1 890
Total operating costs		3 116	3 353	3 400	3 500	3 500	3 500	3 500	3 500	3 500
Taxation	0	0	0	0	0	0	2 822	3 022	3 381	3 421
Total costs	25 850	3 116	3 353	3 400	3 500	3 500	6 322	6 522	8 731	3 921
Revenue	0	6 570	8 213	9 855	10 950	10 950	10 950	10 950	10 950	10 950
Net cash flow	**−25 850**	**3 454**	**4 860**	**6 455**	**7 450**	**7 450**	**4 628**	**4 428**	**2 219**	**7 029**
Internal rate of return (%)	14.39									
Net present value at 12%	2 443									

[a] Replacement of plant and equipment.
[b] Repayment of working capital.

Table 15.8 Calculation of return to equity (in '000 $).

	Year 0	Year 1	Year 2	Year 3	Year 4	Year 5	Year 6	Year 7	Year 8	Year 9
Capital cost	25 850	0	0	0	0	0	0	0	1 850	−3 000
Operating costs		3 116	3 353	3 400	3 500	3 500	3 500	3 500	3 500	3 500
Debt service	1 625	2 308	2 308	5 055	4 726	4 396	4 066	3 737	3 407	3 077
Taxation	0	0	0	0	0	0	2 822	3 022	3 381	3 421
Total outflow	27 475	5 424	5 661	8 455	8 226	7 896	10 389	10 258	12 138	6 998
Revenue	0	6 570	8 213	9 855	10 950	10 950	10 950	10 950	10 950	10 950
Loans	19 233									
Total inflow	19 233	6 570	8 213	9 855	10 950	10 950	10 950	10 950	10 950	10 950
Net cash flow	**−8 243**	**1 146**	**2 552**	**1 400**	**2 724**	**3 054**	**561**	**692**	**−1 188**	**3 952**
Internal rate of return (%)	14.70									
Net present value at 12%	818									

Table 15.9 The financial plan (in '000 $).

	Project	**Increment**
Investment cost	25 850	6 750
Interest during construction	1 525	0
Total	27 375	6 750
Loan	19 233 (70%)	6 750 (100%)
Equity	8 243 (30%)	0
Interest rate 12%		

project. The corporation would, of course, be interested in the contribution of the project to overheads, and if it were necessary to *rank* projects it would choose those making a greater rather than a lesser contribution to overheads. This means exactly the same as saying that it would choose projects with a larger rather than a smaller net present value calculated at the opportunity cost of capital (the discount rate).

Depreciation

Depreciation is ignored in discounted cash flow (DCF) calculations. Depreciation is an accounting concept designed to introduce a measure of the loss of value of physical assets into indicators of corporate performance. Thus, the decline in the value of machinery is a cost to a business because the value of its assets is falling. It does not correspond to real resources moving in and out of the project, although rapidly depreciating assets might lead an enterprise to conclude that some spending on new machinery or equipment or extensive repairs and maintenance were required. It would be the spending on the new machinery that would appear in the cash flow – not the decline in the value of assets.

Working capital

Working capital has been discussed already under 'Investment Schedule'. The rationale for adding back working capital into the revenue stream at the end of the project life is explained under this heading.

The life of the project

The life of this project is shown to be 10 years. For convenience it is often assumed that the life of a project is the same as the expected physical life of the assets. However, in the real world, the period of time over which a piece of equipment may serve some useful purpose is highly variable. It may be in perfect working order but be totally obsolete within a very short period, or its life may be prolonged for years. In practice, physical assets come to the end of their useful lives when they no longer have any *economic* value. In complex projects, especially

Table 15.10 Loan schedule (in '000 $).

	Year 0	Year 1	Year 2	Year 3	Year 4	Year 5	Year 6	Year 7	Year 8	Year 9
Loan	19 233	19 233	19 233	16 485	13 738	10 990	8243	5495	2748	0
Repayments	0	0	0	2748	2748	2748	2748	2748	2748	2748
Interest	1625	2308	2308	2308	1978	1649	1319	989	659	330
Total interest and loan repayment	1625	2308	2308	5055	4726	4396	4066	3 737	3407	3077

where there are alternative means of achieving the same objectives, it will be important to explore various assumptions about project termination. The life of a project is in fact the period that maximises its NPV. To choose a different life from the NPV-maximising life is to opt for a sub-optimal version of the project.

The opportunity cost of capital

The cash flow in this example has been discounted at a rate of 12% per annum. This is the rate of interest that could be obtained by investing the funds in their next best alternative. The opportunity cost is the loss of interest in this next best alternative. If the project fails to generate a positive NPV at this rate, the funds would be better placed in the alternative option.

Risk and uncertainty

The usual approach to risk and uncertainty is to apply sensitivity analysis to the critical project variables. For example, if the predicted catch rate was considered to be particularly uncertain it would be varied, and then the consequences for NPV and the internal rate of return (IRR) assessed. Similarly, if there was uncertainty about shrimp prices in this case, the consequences for the project could be measured. Thus, if the realised price of shrimp falls to $6000 per tonne, the financial IRR falls to 11.5% and the NPV calculated at the discount rate of 12% is –$483 000. We can also easily calculate that the shrimp price at which the NPV at 12% switches from positive to negative (the switching value) is $6330 per tonne.

It has to be recognised that the simple manipulation of data in this way is not an entirely satisfactory way of handling risk and uncertainty, not least because it tells us nothing about the probability of variables diverging from the predicted levels. For example, in a capture fishery project there may be a great deal of uncertainty about the catch rate, but in an aquaculture project it may be possible to reach a well-founded judgement about the probability distribution of different levels of output. Sensitivity analysis does nothing in itself to guide project planners towards a rational appraisal of the risk and uncertainty of different options. The business may be risk averse, in which case it may choose the aquaculture project even if its expected NPV is less than the capture fishery project. On the other hand, risk neutrality on the part of the decision maker may guide it in the opposite direction.

The question of the quantitative assessment of risk and uncertainty is quite complex and is beyond the scope of this book. However, even if a quantitative assessment is not possible, it is most important that the project planner makes an objective qualitative assessment of the project risks.

Taxation

Taxation, calculated at 60.5% of project profits, has been included from Year 6. Until then it is assumed that the project enjoys a tax holiday.

Project financing

Tables 15.9 and 15.10 set out the financial plan for the project. Table 15.9 sets out exactly what must be funded during the construction phase. It is the cost of the investment plus interest payments due in Year 0. Table 15.10 is a repayment schedule for the project. It is a rather rigorous repayment schedule as the loan is totally repaid by the end of Year 9. If the opportunity cost of capital is 12% and the internal rate of return is equal to or greater than this, then the project should at least be able to meet its interest obligations.

Variability of cash flow between different years may mean that, in practice, interest payments to the bank fluctuate. There may be no necessity to repay the loan as well as meeting the interest payments. The business may have other things it wishes to do with surpluses, apart form repaying the principal. On the other hand, if the bank regards the project as risky and insecure, it may want to see its depositors' funds returned and the project may be required to repay quickly so that funds are released for other purposes. Obviously all this will be a matter for negotiation at the outset and probably during the project as well.

In Table 15.10 the loan is repaid on a straight line basis from Year 3. Interest is charged on the amount of loan outstanding at the end of the previous year. This is a simplification that is unlikely to be realised in practice as the bank will charge interest on any outstanding loan; however, the cash flow is usually prepared like this as a matter of convention. It would be possible to amend the schedule to take this into account but at the cost of increasing the complications of computation. The remaining tables (15.11 to 15.14) present the data and computations in a conventional commercial format to complete the picture. Table 15.15 estimates the IRR for the addition of two vessels to the project.

Summary

Valuing projects

The arithmetic of present value

Financial analysis treats the project as a business proposition and assesses whether or not it adds to the net worth of the business. If the net present value (NPV) of the project is expected to be positive, then from the point of view of the business it should be undertaken because it will add to the value of the company. Financial decisions have effects that are spread through time. The effects of an investment decision are a stream of positive or negative cash flows. The money the business spends on the vessels or processing plant or ponds, the operating costs, such as repairs, maintenance and replacement and wages, and the money it will receive for the output from the investment are the cash flows. Money has a time value known as the rate of interest, that is, the rate at which it can be borrowed or lent. This rate is used to *discount*, or reduce the value to the business

Table 15.11 Depreciation schedule (in '000 $).

	Year 0	Year 1	Year 2	Year 3	Year 4	Year 5	Year 6	Year 7	Year 8	Year 9
Buildings and jetty										
Cost	6 000									
Depreciation rate (%)	5									
Depreciation	0	300	300	300	300	300	300	300	300	300
Book value	6 000	5 700	5 400	5 100	4 800	4 500	4 200	3 900	3 600	3 300
Plant and equipment										
Cost	1 850								1 850	
Depreciation rate (%)	14									
Depreciation	0	264	264	264	264	264	264	264	0	264
Book value	1 850	1 586	1 322	1 058	794	530	266	2	1 850	1 586
Vessels										
Cost	13 550									
Depreciation rate (%)	7									
Depreciation	0	902	902	902	902	902	902	902	902	902
Book value	13 550	12 648	11 746	10 844	9 941	9 039	8 137	7 235	6 333	5 431
Total depreciation	**0**	**1 466**	**1 466**	**1 466**	**1 466**	**1 466**	**1 466**	**1 466**	**1 202**	**1 466**
Accumulated depreciation	0	1 466	2 932	4 399	5 865	7 331	8 797	10 263	11 465	12 932
Book value of assets	**21 400**	**19 934**	**18 468**	**17 001**	**15 535**	**14 069**	**12 603**	**11 137**	**11 783**	**10 317**

Table 15.12 Profit and loss account (in '000 $).

	Year 1	Year 2	Year 3	Year 4	Year 5	Year 6	Year 7	Year 8	Year 9
Income									
Fish and ice sales	6 570	8 213	9 855	10 950	10 950	10 950	10 950	10 950	10 950
Expenditure									
Wages and salaries	695	705	715	715	715	715	715	715	715
Utilities	36	43	50	50	50	50	50	50	50
Repairs and maintenance	50	70	100	100	100	100	100	100	100
Fuel	1 000	1 000	1 000	1 000	1 000	1 000	1 000	1 000	1 000
Packaging material	100	200	300	400	400	400	400	400	400
Sales costs	10	10	10	10	10	10	10	10	10
Legal and audit	5	5	5	5	5	5	5	5	5
Licences and royalties	10	10	10	10	10	10	10	10	10
Administration	10	10	10	10	10	10	10	10	10
Travel	100	100	100	100	100	100	100	100	100
Insurance	1 000	1 000	1 000	1 000	1 000	1 000	1 000	1 000	1 000
Project management	100	100	0	0	0	0	0	0	0
Housing maintenance	0	100	100	100	100	100	100	100	100
Total operating costs	**3 116**	**3 353**	**3 400**	**3 500**	**3 500**	**3 500**	**3 500**	**3 500**	**3 500**
Total operating profit	**3 454**	**4 860**	**6 455**	**7 450**	**7 450**	**7 450**	**7 450**	**7 450**	**7 450**
Loan interest	2 308	2 308	2 308	1 978	1 649	1 319	989	659	330
Depreciation	1 466	1 466	1 466	1 466	1 466	1 466	1 466	1 202	1 466
Total other costs	3 774	3 774	3 774	3 444	3 115	2 785	2 455	1 862	1 796
Profit before tax	−320	1 085	2 681	4 006	4 335	4 665	4 995	5 588	5 654
Corporation tax (%)	0	0	0	0	0	61	61	61	61
Corporation tax (value)	0	0	0	0	0	2 822	3 022	3 381	3 421
Profit after tax	**−320**	**1 085**	**2 681**	**4 006**	**4 335**	**1 843**	**1 973**	**2 207**	**2 233**

Table 15.13 Sources and application of funds (in '000 $).

	Year 0	Year 1	Year 2	Year 3	Year 4	Year 5	Year 6	Year 7	Year 8	Year 9
(1) *Net cash flow: real sources*										
Profits before interest and tax	0	987.8	2393.3	3913.8	5983.8	5983.8	5983.8	5983.8	6247.9	5983.8
Depreciation and amortisation[a]	0	2466.2	2466.2	2541.2	1466.2	1466.2	1466.2	1466.2	1202.1	-466.2
Total	0	3454	4859.5	6455	7450	7450	7450	7450	7450	7450
Uses										
Fixed assets	21 400									
Other capitalised investment	1450								1850	-3000
Total	22 850								1850	-3000
Cash flow: real	**-22 850**	**3454**	**4859.5**	**6455**	**7450**	**7450**	**7450**	**7450**	**5600**	**10 450**
(2) *Net cash flow: financial sources*										
Issued share capital	8242.5									
Loan	19 232.5									
Total	27 475									
Uses										
Debt service	1625	2307.9	2307.9	5055.4	4725.7	4396	4066.3	3736.6	3406.9	3077.2
Taxes	0	0	0	0	0	0	2822.3	3021.8	3381.0	3420.7
Other	3000	1146.1	2551.6	1399.6	2724.3	3054	561.4	691.6	-1187.9	3952.0
Total	4625	3454	4859.5	6455	7450	7450	7450	7450	5600	10 450
Cash flow: financial	**22 850**	**-3454**	**-4859.5**	**-6455**	**-7450**	**-7450**	**-7450**	**-7450**	**-5600**	**-10 450**

[a] An annual cost against project revenue to offset pre-operating expenses (£1450 000) and interest incurred during construction (£1625 000).

Table 15.14 Balance sheet estimates (in '000 $).

	Year 0	Year 1	Year 2	Year 3
1 Assets				
1.1 Current – total	3000	4146	6698	8427
1.1.1 Cash	1500	2646	5198	6927
1.1.2 Receivables	1500	1500	1500	1500
1.1.3 Inventories	0	0	0	0
1.2 Fixed (net) – total	21 400	19 934	18 468	17 001
1.2.1 At cost	21 400	21 400	21 400	21 400
1.2.2 Accumulated depreciation	0	1466	2932	4399
1.3 Land	0	0	0	0
1.4 Buildings (net)	6000	5700	5400	5100
1.4.1 At cost	6000	6000	6000	6000
1.4.2 Accumulated depreciation	0	300	600	900
1.5 Equipment (net)	15 400	14 234	13 068	11 901
1.5.1 At cost	15 400	15 400	15 400	15 400
1.5.2 Accumulated depreciation	0	1166	2332	3499
1.6 Other capitalised assets	3075	2075	1075	0
1.6.1 At cost	3075	3075	3075	3075
1.6.2 Accumulated amortisation	0	1000	2000	3075
1.7 Other (securities, etc.)	0	0	0	0
Total assets	27 475	26 155	26 241	25 428
2 Liabilities	19 232.5	19 233	19 233	16 485
2.1 Current	19 232.5	19 233	19 233	16 485
2.1.1 Payables	0	0	0	0
2.1.1.1 Suppliers	0	0	0	0
2.1.1.2 Banks	0	0	0	0
2.1.2 Current long-term debt	19 232.5	19 233	19 233	16 485
2.1.3 Other	0	0	0	0
2.2 Long-term	0	0	0	0
3 Net worth	8242.5	6922	7009	8614
3.1 Paid in equity	8242.5	8243	8243	8243
3.2 Retained earnings	0	– 1320	– 1234	371
3.3 Other	0	0	0	0
Liabilities and net worth	27 475	26 155	26 241	25 099

of the future stream of net cash flows. If the net present value at the relevant discount rate is positive, the project should go ahead.

Internal rate of return

The internal rate of return (IRR) is the rate at which net benefits are discounted to give an NPV of exactly zero.

Cost–benefit ratio

The cost–benefit ratio is the discounted stream of benefits divided by the

Table 15.15 Calculation of impact of incremental vessels (in '000 $).

	Year 0	Year 1	Year 2	Year 3	Year 4	Year 5	Year 6	Year 7	Year 8	Year 9
Capital costs (two vessels)	6750	0	0	0	0	0	0	0	0	0
Operating costs		1657	1657	1657	1657	1657	1657	1657	1657	1657
Revenue		2370	2962.5	3555	3950	3950	3950	3950	3950	3950
Net cash flow	−6750	713	1305.5	1898	2293	2293	2293	2293	2293	2293
IRR (%)	21									

discounted stream of costs. If the ratio is greater than one this implies that the gross benefits exceed the gross costs, or that the NPV is greater than zero. The decision rule would be to accept projects with a positive cost–benefit ratio.

Payback criterion

The payback rule is that projects are ranked according to the payback period. Those whose positive cash flow meets the cost of the investment in the shortest time are the projects to be accepted.

Accounting rate of return

The accounting rate of return (ARR) is the accounting profit of the project divided by the book value of the assets employed in it.

Valuing projects under uncertainty

Authorities usually recommend 'sensitivity analysis' as a means of handling the issue of uncertainty. This means simply trying out plausible alternative values for the variables in the appraisal and seeing what the effects are on the project as a whole. We see how sensitive the project is to changed assumptions.

Planning of capital structures

The capital structure of the project is the ratios and amounts of each type of funding used to pay for it.

Chapter 16

Development of the Fish Business in Vanuatu: A Case Study

Introduction

The following case study is based on a project evaluation prepared for the European Development Fund. The evaluation was prepared by the author and Mr Richard Stride (Palfreman and Stride, 1996). It is adapted here to illustrate the problems and opportunities in private sector fisheries development. Special emphasis is placed on identification of distinctive capabilities (architecture, reputation, innovation and strategic assets) within the sector and building upon them. This adaptation is reproduced with the permission of the Government of Vanuatu.

The Fisheries Extension Service and Training Centre Project was identified during the latter years of the Village Fisheries Development Programme (VFDP), round about 1985. The VFDP which had run from 1982 was thought to be becoming weaker because the relationship between inputs and outputs at the basic level was not being managed carefully enough and was susceptible to improvement through training and extension. It was thought that providing local back-up in the villages, to help people maintain their vessels and engines and to market fish locally or nationally, would encourage people to fish when they might have otherwise chosen to do something else with their time. Hence the project: a set of funded measures from 1987 to 1995 to support the development of a fishing industry.

The project was intended to improve the *architecture* of the industry through training and extension – encouraging people to work more efficiently and to market fish better. It was also intended to encourage *innovation* – to use new and improved technologies in catching and distribution. Did it succeed? The project met its objectives within time and budget, but would have been strengthened with a greater understanding of the fisheries economy and the commercial motives of the main players.

Project identification, preparation and design

Background

The Republic of Vanuatu consists of an irregular archipelago of about 80 islands in the south-west Pacific Ocean, lying about 1000 km off Fiji and 400 km north-east of

New Caledonia. The group extends over a distance of about 900 km from north to south. The islands have an oceanic tropical climate with a season of south-east trade winds between May and October. Winds are variable, with occasional cyclones, for the rest of the year. Annual rainfall varies between 2300 mm (90 in) in the south and 3900 mm (154 in) in the north. In the capital, Port Vila, the mean temperatures vary between 22 and 27°C.

Identification and formulation

The project under discussion was identified during the latter years of the Village Fisheries Development Programme (VFDP). The Second National Development Plan or DP2 (Republic of Vanuatu, 1986) included a clear statement of requirement for the training centre and an extension service for the fisheries sector.

It seems that there was a perception that the VFDP was becoming weaker at this time. DP2 referred to a fall in the mean volume of fish per village project from nearly 5 tonnes in 1983 to under 2 in 1986. Others (Shepard, 1988) also observed rising numbers of business failures and the decline appeared to continue. A file note in 1988 referred to the difficulties of securing bait, unfavourable weather conditions, poor equipment, the use of fishing vessels as taxi boats, a shortage of inputs and marketing difficulties. David and Cillauren (1992) reported that the decline in tonnes caught per village project continued into 1988 (Table 16.1).

Table 16.1   Development of fisheries production supported by the VFDP.

	1983	1984	1985	1986	1987	1988
Number of village projects	11	23	50	72	59	75
Total reported catch (tonnes)	49.1	87.9	97.5	128.9	93.5	79.3
Average annual production per village project (tonnes)	4.5	3.8	2.0	1.8	1.6	1.1

Source: David and Cillauren (1992).

It seemed a logical argument. The VFDP was thought to be growing weaker because the relationship between inputs and outputs at the basic level, the 'production function', was not being managed carefully enough and was susceptible to improvement through training and extension. It was thought that providing local back-up in the villages, to help people maintain their vessels and engines and to market fish locally or nationally, would encourage people to fish when they might have otherwise chosen to do something else with their time. For these reasons the Integrated Training and Support Complex and the establishment of the Extension Service were included in the government's plans.

Importance of thorough *economic* analysis at an early stage

Three issues illustrate the importance of a thorough *economic* analysis of the development process at an early stage – at project identification:

(1) Through the history of the VFDP and subsequently the Extension and Training Project there was never a recognition that the basic *economic* case for project intervention is the correction of market failure. Economic theory states that in the absence of an identified market failure, the market of itself will supply the incentives for the development of the private sector in fisheries. The key obstacles to further development of the sector were (a) relatively high unit costs of catching and distribution and (b) market prices which were not high enough in more remote areas to compensate for the high production costs. Deepening of the level of skills through training and extension – that is, trying to influence what has been termed the architecture of the industry – was not obviously the best response to these fundamental economic constraints.

(2) The DP2 referred to periods when the national state-owned fish market, Port Vila Fisheries Limited, was obliged to restrict the quantity of fish purchased because of limited storage capacity and financial reserves, thus reducing the average value received by fishermen. Lindley (1993) also refers to a fall in the real price of fish when he arrived in 1989. The economic point is that training and extension might reasonably have been expected to influence the functional relationship between inputs and output of fish, but the place along the production function where fishermen choose to apply their inputs and produce fish depends on the economic environment in which they operate, notably prices received and their alternative sources of income and other sources of utility. There is, actually, nothing untoward about low levels of fish production if that level of fish production is a result of transparent market forces. A logical economic aim is for any unnecessary inefficiencies, such as information failures, problems caused by overfishing or savings gaps to be reduced or eliminated.

(3) The third economic point that was not considered at the project identification stage was the relationship between the size of the sector and the public sector resources which it was proposed should be used to support it. The project was in danger of soaking up all the economic added value available from fisheries rather than creating more.

The role of the European Union

The project was financed by the European Union's Development Fund (EDF). The Financing Proposal (CEC, 1987b) was approved by the EDF on 21 September 1987 and signed as a Financing Agreement (CEC, 1987a) on 25 September 1987. The purpose of the project was (as set out in these official documents):

(1) To establish an Extension Service to serve the developing small fishing industry throughout the archipelago of Vanuatu. Four centres were to be established under the project. At each centre an expatriate extension adviser and a Ni-Vanuatu counterpart were to conduct training and provide advice to all fishing enterprises in the area.

(2) To provide for a fisheries training and support centre in Luganville, the geographical centre of most fishing operations and vessel research in the country, to be operated by the Fisheries Department. The first activity of the training centre was to be the training of new fisheries extension officers.

EU records indicated expenditure of about 1.6 million ecu compared to a total budget of 1.7 million ecu at the time of the study. The Vanuatu government's contribution was in kind and salaries. Approximately 1 million ecu was spent on the physical investment, leaving 0.6 million ecu for associated work programmes. Other donors have also contributed to the project.

Project management

Merits of keeping projects simple

In principle it would have been preferable to have planned, financed and administered the training centre and the extension support as separate projects. The reason for this is that the two components of the project each had a distinct project purpose which required separate project management and distinct measures of achievement. For example, seeing the components as separate might have pointed planners to linking the training component of the project to other, non-fisheries, training endeavours in Vanuatu – but this was not, apparently, explored.

Project administration

The project required a series of initiatives and decisions, such as the recruitment and management of expatriate village extension advisers, which the government administration was not ready for. This resulted in a larger role for the expatriate senior project advisers than was originally intended or desirable. In particular, because expatriate advisers do not stay indefinitely, the chances of sustainability are reduced. When they go there is a danger that they leave a vacuum, unless local staff are fully prepared to take over, and therefore the risk of collapse of what the project has achieved.

The role of cost–benefit analysis

Cost–benefit analysis of projects (Chapters 14 and 15) has tended to fall into disrepute, basically because it has been perceived as a rubber stamp rather than an objective project test. There is still a case for undertaking cost–benefit analysis of projects, although none was done in this case. A cost–benefit analysis might have suggested that the scale of the project as designed was too large for any reasonable prediction of expected benefits, and would have encouraged planners to look towards a lower cost means of obtaining the expected benefits.

Use of a logical project framework

Ideally, projects should be managed under the guidance of a clear project strategy summarised in the logical project framework (logframe). A logframe would have provided, on single sheets of paper, a commonly understood and uncomplicated statement of the purposes of the two components of the project, together with the indicators that it was intended to observe and record as an assessment of project achievements.

Other project management practices

Project management would have been strengthened by a simple statement of the main principles and policies guiding the project. Areas where a clear statement of principles, pulled out from the Financing Agreement, would have been helpful include the following:

- A project 'mission statement' defining the project purposes. Everyone concerned is then aware what they are there for and when the project purposes are achieved.
- Assignment of responsibility for the project to the Director of Fisheries. It would not, of course, prevent the Director of Fisheries from delegating responsibility, but *single point responsibility* is an important constituent of project success.
- The responsibilities of expatriate advisers in relation to the project, including the programmed transfer of skills and responsibilities to counterpart staff.
- The responsibilities of the EU, other donors and the government of Vanuatu.
- Responsibility for project monitoring, including regular formal discussion of the results of monitoring.
- Change management procedures – what happens if the project policies and principles are not met, if the logframe becomes irrelevant and if the specified milestones are not achieved.
- A milestone plan for each of the two components (training centre and extension support). This would have organised the project results into a sequence, leading logically through to the achievement of the purposes, which could

then have been subject to routine monitoring by senior management and advisers. Examples of appropriate milestones for the Training Centre might have been 'When the classrooms are completed', 'When local staff are appointed', 'When timetables are prepared' and, crucially, 'When the Ni-Vanuatu manager takes on full responsibility for the operation of the Fisheries Training Centre'. Examples of appropriate milestones for the Extension Service component might have been 'When the Extension Centres are constructed', 'When the expatriate advisers' contracts are completed', 'When the Extension staff are appointed', 'When the Extension Centres are sold', and, crucially, 'When the government of Vanuatu takes on full financial responsibility for the operation of the Extension Service'.

Financial management

Both components of the project have been involved in semi-commercial activities. The Extension Service workshop bought and sold equipment and supplied commercial services. In the case of the Training Centre the main commercial activity is catching, processing and selling fish, which also uses a revolving fund. The project review team concluded that the Vanuatu Fisheries Department staff, within the limits of the public accounting system, kept very good track of the commercial operations of the project. However, true commercial accounting, taking account of changes in asset values, maintaining excellent books and presenting management accounting reports to a decision-making group was completely absent.

Relevance of the project

The Economy

The 1993 estimate of GDP was $181 million, or $1131 per capita (164 000 people). At the time of the study, GDP was rising, after a period of slow growth in the 1980s, and was estimated to have increased by about 5% between 1992 and 1993; however, the impact of increased national income on living standards was weakened by a population that was growing rapidly, at about 2.9% per year between 1989 and 1994, such that real per capita incomes were tending to decline. Quite evidently, there is a real social need for economic activities that employ and provide incomes for the Ni Vanuatu people.

At the time of the census in 1989 about 61% of the employed labour force was engaged in agricultural activities. Copra, beef, cocoa and timber are produced largely for export. Various other garden produce, such as taro and yams, is produced for subsistence. The Vanuatu government has encouraged the development of a forestry industry. The industrial sector contributes about 12% of GDP, although only 3.5% of the labour force is employed in it. Manufacturing does take place, mainly for processing agricultural products. Vanuatu's first plywood manufacturing mill was opened in Santo in 1991. The major expansion over the last 10 years has been in the service sector, which now contributes about two-thirds of GDP. Tourism, 'offshore' banking facilities and an open shipping registry all make a contribution to the country's national income. Vanuatu continues to run a substantial deficit in trade in merchandise, but this converts to a current account surplus when aid inflows and returns to tourism are taken into account.

This project took place under the guidance of the Second National Development Plan (1986–1991). Development objectives for the fisheries sector as specified in the Plan were as follows:

• Develop the exploitation of fisheries resources to achieve its potential as an important economic activity.
• Maximise the sector's contribution to an expansion in the nation's income-earning and employment opportunities.
• Increase the production of fish and other marine products for domestic and overseas markets.
• Reduce the level of canned and fresh fish imports.
• Increase the sector's contribution to government income available to support other areas of social and economic development.

These objectives were to be achieved within the framework of the overall National Development Objectives. The principal strategies aimed at achieving these development objectives were:

• The establishment of a decentralised Fisheries Extension Service.
• Upgrading of training, boat and engine repair and maintenance, marketing, gear supply and other related departmental support activities.
• Building of a new integrated training and fisheries support complex at Luganville (the main town on Espirtu Santo).
• Increasing the number of established staff within the Fisheries Department.

The project continued into the Third National Development Plan (Republic of Vanuatu, 1992). This document reflects more modest aspirations for the fisheries sector, and explicitly recognises poor returns to fishing vessels as being a major weakness in fisheries development. The development objective for the sector now was to continue to encourage and assist the fisheries sector to promote fishing as an economic activity, and expand levels of fishing activity, marketing and processing.

The following strategies were recommended:

• Rationalise fisheries extension centres based on levels of fishing activity and privatisation opportunities for ice-making facilities.

- Continue training of fishermen.
- Privatise PVFL (Port Vila Fisheries Limited – the state-owned marketing company) fish markets.
- Reactivate and expand SPFC (South Pacific Fishing Company) to include fish processing.
- Upgrade boat construction facilities at Luganville (main town on Santo) with an eventual view towards privatisation.
- Continue fisheries research through the auspices of ORSTOM (French-supported research body).

Fish resources of the reef slope

Since independence in 1980, the government has supported fisheries development. Given the volcanic character of the islands in an ocean of deep, unproductive water, any visions that may have been harboured of a prolific maritime zone were always bound to be disappointing. Nevertheless, biological studies on the deep-water resources led to the conclusion that there might be commercially viable fish species available on the reef slope and, in addition, it was known that pelagic fisheries resources were available, at least seasonally. From August 1980 to June 1981 the South Pacific Commission carried out test-fishing for demersal fish with deep-water drop lines and achieved catch rates that were believed to be commercially viable. It was, therefore, a logical step to initiate a period of development. These resources (as opposed to reef and oceanic) have been the core of the government's fisheries development efforts to date.

The first attempt to quantify the fish population dynamics of the deep-water species was provided by Brouard and Grandperrin (1984). *Les Poissons Profonds de la Pente Récifale Externe à Vanuatu* brought together the conclusions of some research on the demersal species of fish inhabiting the reef slope of the Vanuatu islands. The fish species are various kinds of snapper, in particular *Etelis* spp. and *Pristopomoides* spp. Brouard and Grandperrin made use of the similarity between the geography of Hawaii and Vanuatu to derive some initial estimates of maximum sustainable yields (MSY).

Other studies have since been completed. Carlot and Cillauren (1990) undertook an analysis of limited data and concluded that the MSY might be, perhaps, between 113.2 and 603.5 tonnes per year. The conclusion is based on a statistically significant fit to a biological production model for two islands, Ambae and Paama. Data were recorded for other islands. On Efate, where fishing effort was also high, like Ambae and Paama, no fit to the model used was found. The general position in Vanuatu, as confirmed by Carlot and Cillauren, is that effort and catch in most areas are positively correlated with a linear regression line. In other words, for the present, a general conclusion of the Carlot and Cillauren study is that the more vessels fish the more they catch. Dalzell and Preston (1992) analysed the results of the activities of the dropline fishery in a number of Pacific islands. The general conclusion is summarised in the following quotation:

'These fisheries will continue to endure in those countries discussed above, but its seems likely that the period of research and development of this method of fishing is now drawing to a close.' (p.33)

The reality is that biological estimates of the fish population dynamics of deep-water fish in Vanuatu are all preliminary and tentative. There are not enough data to make any certain statement about the deep-water stocks. The situation of uncertainty among the biological parameters is typical of developing countries, and by no means unknown in the developed world. The question for the government of Vanuatu is, therefore, how it is going to *manage* this uncertainty. There are four distinct arguments, all of which point away from more biological research.

(1) The fishery is a high cost fishery and therefore as soon as catch rates begin to be affected by declining fish stocks the costs of fishing will discourage further fishing effort.
(2) It is possible to estimate by means of a bioeconomic model the maximum possible surpluses (the main source of EAV for the fishermen) from the fishery and they are certain to be small, under any reasonable assumptions, perhaps in the order of vt (Vatu) 6 million for Efate, vt 6 million for Santo and vt 4.5 million or less for Malekula, with much smaller or negligible surpluses for other islands. Under these circumstances it is not likely to be worthwhile for the government to spend scarce public sector resources on attempting to improve understanding of the fish population dynamics of the fishery and then to attempt to manage it with the means at its disposal. The return to the national economy in the form of economic rents does not warrant it. Because the costs of fishing are so high and the price of fish, net of transport costs, is so low, extracting EAV from the resource is not easy.
(3) Most fishing for deep-water species takes place on the reef slope, within the 6-mile management domain of the Provinces. If catch rates do begin to fall in the areas close to the shore, Provincial Councils, in cooperation with local chiefs, can implement informal or formal regulations to reduce fishing effort.
(4) In any event, because the deep-water fish are slow-growing, there is a case for regarding them as non-renewable resources, to be exploited and then left to recover – a 'pulse fishing strategy'. It is reported that some Santo fishermen have been doing this, which may make economic sense.

Extension advisers in the villages

The problem of falling catch rates was attributed (incorrectly) only to technical deficiencies. Thus, it was argued, if only skill levels were improved (better outboard and vessel maintenance, fish handling, fishing practices, etc.) production would rise. However, the socio-economic context of falling catch rates was not analysed. If the real price of fish is not high enough or if the costs of fishing are too high, as indeed they were, then no amount of technical adjustment to the production function will induce people to fish more.

The role of the expatriate advisers was not considered carefully enough. Unless people in the villages actually want advice every working day there is simply not enough for the adviser to do. The jobs were not full-time occupations. A better model would have been for the Fisheries Department to be available to advise if requested and to make periodic visits to fishermen to check that certain procedures were being followed – but essentially to stand back and let potential commercial fishermen make their own decisions, seeking advice when they require it.

Extension initiatives during the project included:

- Canoe Modernisation Fund – a revolving fund for supplying small items of equipment. It was terminated because of difficulties of administration.
- Fish Aggregating Devices – deployed for the benefit of fishermen, but the results were inconclusive.
- Lease Scheme – selected fishermen were allocated a vessel which they used to fish commercially under close supervision. The income was shared between the lease fee, trip operating costs, fisherman's income and savings. The scheme was brought to an end through administrative difficulties, rather than a failure of principle.
- Alternative Methods of Fish Preservation – there were some trials, but most accept there is no significant market for preserved fish in Vanuatu.
- Fishery Support Services – the main focus of the project was the provision of support to the mechanised deep bottom fishery. Activities included ice production, fish marketing, training, engine and boat repair and maintenance, sale of fuel and fishing inputs, location of fishing grounds, experimental fishing and data collection and the administration of the lease scheme.

Each of the rural extension centres was manned by an Extension Officer who was assisted and trained by an expatriate Extension Adviser. A carpenter and village mechanic were trained at Luganville and maintained engines and boats using the centres' simple workshop facilities.

A process of withdrawal by the public sector from commercial activities at the main centres began in 1992. The new private sector operators are fishermen or boat owners who are able to derive both direct and indirect benefits from the operation of the markets. However, the project continued to subsidise the maintenance of ice plants through the operation of the mobile workshop and did pay the cost of certain repairs made necessary by the age and condition of the machines at handover. At the end of the day, however, the withdrawal of project support will most probably imply the end of the centres.

The Fisheries Training Centre

The Fisheries Training Centre (FTC) was intended to provide training to the seven Extension Officers and to fishermen wishing to enter the small-scale commercial fishery and those already engaged in it. Initially, training was carried out in improvised conditions in a disused shell factory, at the site, moving to the purpose-built accommodation in February 1990.

The development of the training courses has been the primary responsibility of the Training Adviser (aid-funded), as envisaged in the Project Proposal, and the Centre Manager (local) in consultation with the fisheries extension staff. There was a continuous process of course development in accordance with the experience gained and with the identification by the Extension Service of further relevant areas for training.

More recently, in connection with the development of methodology for the reef management enhancement programme, a training workshop for Extension Officers was held at the FTC. David's research (1988 and 1987) and David and Cillauren (1992) were useful foundations for this. This was the first of a series of workshops which led to the compilation of traditional and scientific knowledge of reef ecology and management and the formulation of appropriate management advice for reef owners.

At the time of the study, the FTC maintained a fleet of four fishing boats for training, fishing and demonstration purposes. Commercial fishing operations were intensified in response to the prospect of falling external funding in 1994. The FTC has made investments in equipment to enhance its production and marketing capacity. The operation of these vessels has provided:

- Training opportunities. The boats are manned by trainees or casually employed fishermen who benefit from the experience.
- The opportunity to assess vessel types introduced by the project.
- A series of accurate catch and effort data for the deep bottom and FAD (fish aggregating device) fisheries.
- Development of fishing vessel management capacity at the centre.

The essence of the FTC problem is low numbers of fishing trainees and the prospect of insufficient income to cover costs.

Project monitoring and evaluation

Project monitoring

The project monitored its physical and financial implementation through annual activity reports. Few specific indicators were specified in the project documents or set by the project. The lack of indicators, particularly the absence of a time frame for project activities, makes it difficult to determine the degree of progress made towards the achievement of objectives through the annual reports alone, although they do give a clear picture of progress against the annual workplans.

Internal monitoring could have been improved. The project needed two robust milestone plans, one for each component of the project. Then the core of the monitoring system would have been a system of checking off each milestone, including those concerned with internal aspects of the project, as it was achieved. The project manager, ultimately the Director of Fisheries, but delegated if appropriate, would then have been responsible for checking that both components of the project are moving in the agreed direction towards the project purpose. It is particularly important to do this with a process project, because it is very easy to become diverted.

Project evaluation

The Shepard report

In 1987 a Canadian consultant, Dr M.P. Shepard, was commissioned to prepare an evaluation of the VFDP. The main thrust of the Shepard recommendations (Shepard, 1988) was towards closer monitoring of the fishery through a computerised system. Shepard was concerned to improve the management of the VFDP and argued that closer monitoring of the performance (catches and commercial performance) was the best means of achieving this. Initially, an attempt to improve record keeping was established, and a residual system of data collection, although not of data analysis, still exists. However, it seems that the Vanuatu institutional setting was not sufficiently robust for these proposals. In making this comment it is worth recalling that data collection from fishermen is notoriously difficult throughout the world. People are reluctant to advertise their successes or failures and see no reason to supply data to government officials. On the government side, the failure of the Extension Service to monitor the progress of the enterprises is less understandable, but perhaps reflects the complexity of the task that was set. In any event, quantitative monitoring of the successor of the VFDP was not achieved.

There was also at this time an EU-funded evaluation of the VFDP (by the North Sea Centre Group, 1987). The recommendations of this report also entered the objectives of the Extension Service, which included 'To carry out the recommendations of the Shepard and North Sea Centre Reports'. It appears that the North Sea Centre consultants were only in Vanuatu for a week which seems an unrealistically short time to review the VFDP and make some useful suggestions for the future.

During 1988, and in possession of the North Sea Centre and Shepard evaluations of the VFDP, the Fisheries Department reassessed the project and decided to reduce the scale of the Extension Service and introduced a number of new elements designed to enhance the prospects for achieving sustainability and for broadening the support given to the subsistence fishery (e.g. the Lease Scheme, Canoe Modernisation Fund and development of vessels with lower capital and running costs).

The mid-term review

The North Sea Centre Group published an Evaluation Report of the project in 1987. It was intended to re-orient the project towards fisheries management, lower cost fishing methods, improved statistical collection and FAD development. A conspicuous recommendation was that expatriate involvement in extension should be drawn to a close. At this point the objectives of the project were largely achieved. The Fisheries Training Centre was running, although the classrooms would only just have come into use, and the Extension Service was operating, and, would operate better without the expatriate contribution, according to the Report. It might have been logical at this point to have drawn the project to a close, and to allow the Fisheries Department to take up its on-going responsibilities. The project lost some of its 'goal-directedness' at this point and became more of a means of supporting on-going Fisheries Department activities.

   In the latter years of the project, expenditure was mainly in support of the Associated Work Programme, in practice mainly Fisheries Department operating expenses (as can be seen from the detailed budget submissions). It would have been preferable to have organised this later expenditure, if it was justified, within the framework of a new project, with its own project purpose and milestone plan, rather than associating it with the old project. There were new issues to be dealt with, such as directing the work of the Extension Service to canoe fisheries and reef management and developing alternative uses for the Training Centre, which might have been addressed more coherently within the framework of a new project.

Risks

Funding from the government of Vanuatu

A major risk with this project was that the government of Vanuatu would be unable to supply the financial resources to continue the activities after the expiry of the external funding. Indeed, expecting the government to be able to do this was an error of judgement at the planning stage.

Deep-water fishing may not be viable

The project was focused on supporting the deep-water fishery. Its exploitation has indeed turned out to be more problematic than was originally supposed. Nevertheless, it is now a real commercial fishery operating most successfully in regions of Vanuatu which have access to the markets which can afford the relatively high cost of catching and distribution.

Declining departmental commitment to the project

Declining departmental commitment to the project was not anticipated, but clearly has been a problem, and also recurring in the shape of a public service strike.

Institutional capacity and culture

Project bookkeeping was done well. No one has offered any adverse criticism of the way that FTC has run the various courses and the fishermen contacted have enjoyed them and feel that they have benefited. The FTC has also been complimented on the way non-fisheries course are serviced. In other respects the administration has been less successful – such as maintaining the statistical system. The weaknesses that have arisen can, in part, be attributed to the strike of public servants in November 1993 which resulted in the dismissal from the service of a number of experienced staff members.

The Fisheries Department requires guidance in certain respects. One has to remember that Vanuatu is a very small country in which institutional capacity is still limited. It is unrealistic to expect local staff to take new initiatives, such as project formulation, without external assistance, but they can reasonably be expected to supply comprehensive administrative support to new initiatives, provided that what is required of people is made clear and is carefully monitored. Projects, therefore, should be kept simple, with limited objectives, to be achieved over a short time period. Back-up support required should be clarified at the outset; this is basically good straightforward project management practice.

Impact

Project purposes and overall objectives

The project successfully established an Extension Service and a Fisheries Training Centre, thus meeting the project purpose. However, the study authors judged that the most important contribution of project was in strengthening human resources.

Contacts made with fishermen suggest that the courses offered by the FTC are well received. In particular, courses in outboard motor maintenance and offshore fishing experience were mentioned. Furthermore, there is evidence that there is still a demand for the Rural Fishing Skills course. At the time of the evaluation the FTC had applications from over 30 individuals on their books. The courses are well-designed.

The impact of the Extension Service is harder to identify because it is more diffuse, but at the time of the study there were eight functioning marketing centres, now being run by private operators who buy fish from fishermen and sell it either locally or in the population centres. Also, the fact that fishermen in Santo, Malekula and Efate are complaining of declining catch rates is indicative of the fact some fishing is taking place. (Incidentally, just because catch rates; e.g. catch per day or per reel hour, are declining is not, in itself, evidence of overfishing. Catch rates of unfished or lightly fish stocks are normally expected to fall as fishing effort increases.) Furthermore, deep-water fishing skills which were rare 10 to 15 years ago are now widespread in Vanuatu. The fact that the vessels are used only part-time is a reflection of rational economic choices, not at all a sign of project failure.

Future developments

The Extension Service

A key aspect of the policy environment is that the government faces continued difficulty in justifying public spending. It follows that the Extension Service has to deliver concrete and visible benefits if the government is to be persuaded to meet its costs. Improving the performance of the public sector was one of the key policy areas identified in the Third National Development Plan (DP3). Fallon (1994) identifies three major areas where progress needs to be made: (1) the policy development, coordination and monitoring process urgently needs strengthening; (2) the management, skills and efficiency of the public sector need to be improved; (3) policies should be oriented towards the improvement of the regulatory environment.

There are also areas where the Fisheries Department can make progress. For the Extension Service, monitoring the marine environment and the development of the fisheries is an important task for the future. It must become the eyes and ears of the Fisheries Department so that ministers and others are fully informed of developments. The Service needs to play its part in the management of Vanuatu's natural resources. The Service must also contribute towards the regulation of the coastal and marine environment. The government has published a statement of environmental policy (Vanuatu National Conservation Strategy, 1993). This includes the goal 'Ensure biological resources are used sustainably'. The Fisheries Department has a part to play in developing environmental awareness and promoting compliance with fisheries regulations.

However, this scenario, in which the Extension Service contributes to the optimal management of the environment, is not one in which it can recover its costs from beneficiaries, so it will have to depend on public sector resources. To be sustainable the Extension Service needs to supply services to Vanuatu that are widely recognised as making a significant contribution to the development of society. It is most important that this point is fully appreciated by the senior officers in the Fisheries Department – they must deliver and be seen to deliver. Only then can the support from the recurrent budget be justified.

The Fisheries Training Centre

The post-project infrastructure of the Training Centre is too large for the needs of the fishing industry. FTC management has been quite successful so far in securing other users of the space. The evaluators believe that the government of Vanuatu recognises that this is a useful asset and will ensure that it is used for the benefit of the country. The FTC in its present form is not sustainable in the long term. The demand for the Rural Fishing Skills course will decline in time and new sources of income will be required. Subsidising the FTC from earnings from fishing is not a practical long-term solution because the FTC vessels cannot reasonably be expected to cover more than their own long-run opportunity costs – in other words, they can only be reasonably expected to earn enough to cover operating costs, depreciation and interest payments. In any event, commercial fishing should be a commercial, private venture rather than a government activity. It would be unreasonable to expect the recurrent budget to continue to meet all the costs of the Centre indefinitely.

The fisheries policy environment

The Fisheries Department needs to reorient itself towards the new commercial realities – it needs a clear statement of policy. The policy statement should define the policies of the country towards the sector. The study recommended that the Fisheries Department should become a *facilitator* rather than initiator of fisheries development, recognising that long-term prosperity of the fishing industry depends on *private sector*, rather than public sector, initiative. To foster this, the Department should emphasise the role of *information* in improving the functioning of the production and distribution system. For example, the Department needs good quality information about the state of fisheries resources and the industry needs good quality information about resources, markets, standards, prices and suitable fishing techniques. If they can obtain these readily, the industry functions better. It is recommended, then, that an improvement in the flow of information should be a fundamental feature of the Department's mission.

The policy statement should be submitted to the Cabinet for ministerial approval. The document would aim to set out the guiding principles of the government of Vanuatu in its management of the fisheries sector. It would then set out the sectoral objectives and strategies. It might also include some possible projects or programmes that contribute towards the objectives and strategies.

The study recommendations

Recommendation 1

The first recommendation follows from the above comments on the policy environment. The Fisheries Department needs to become a facilitator of private sector development.

The Fisheries Department should prepare a statement of policy setting out guiding principles, objectives and strategies for the management and development of the fisheries sector. The document should also outline a selection of projects and programmes designed to contribute to the defined strategies.

Some aspects of the work of the Fisheries Department need to be rebuilt. For example, the system for the collection and analysis of statistics needs re-establishing and the Extension Service needs guidance on the formulation of its work programme. The Department also needs to rethink its approach to fisheries management and development.

Recommendation 2

The Government of Vanuatu should have the benefit of an adviser to the Director of Fisheries at a senior level, designated Adviser on Fisheries Policy to the Director of Fisheries, and reporting directly to the Minister responsible for fisheries.

The adviser should assist in the work of policy, programme and project formulation and management. To achieve this, he should have access to short-term technical assistance for specialist inputs and should report and have regular access to the minister responsible for fisheries. The adviser's office should be in the Fisheries Department in Port Vila. Duties should include:

- Advising the Director of Fisheries in the preperation of an official policy statement.
- Assisting the Director of Fisheries in the preparation of a work programme for the Department, based on the policy statement.
- Assisting the Director of Fisheries in the implementation, monitoring and evaluation of the work programme.
- Advising the Director of Fisheries in the formulation of a training programme for the Fisheries Department.
- Advising the Director of Fisheries in the development of a statistical system which meets reasonable government requirements.
- Advising the Director of Fisheries in all other areas of policy and management as may be required.

Recommendation 3

Under technical assistance, a Commercial and Accounting Adviser should be appointed to advise the Director of Fisheries on the planning and implementation of commercial bookkeeping and accounting practices at the Fisheries Department in Santo as well as providing a service equivalent to a company secretary for the semi-commercial activities of the Department.

There is a separate requirement for disciplined commercial accounting in the three semi-commercial activities in Santo. In support of this recommendation, for example, Table 16.2 is an assumed income and expenditure account for the Fisheries Workshop in Santo. The Fisheries Workshop was established with aid funding as a semi-commercial operation to support village projects. As soon as costs, currently met out of the recurrent budget, are included a surplus turns into a deficit. This is not surprising because the Workshop was not operating under commercial constraints. However, in the future, when finance becomes tighter, strict financial management will become essential.

The evaluation mission recommended that the Director of Fisheries should have the benefit of a Commercial and Accounting Adviser, based in Luganville, who would also act as a company secretary. The Adviser should report to the Fisheries Policy Adviser.

He/she should have the following responsibilities:

- To prepare accounting systems for (1) the Fisheries Training Centre, (2) the Fisheries Workshop and (3) the boatyard. The accounting systems should be designed to ensure that the Director of Fisheries has access to regular accounting information which fully reflects the commercial performance of the three entities. The systems should also ensure that control is exercised over cash transactions.
- To advise the managers of the three areas of commercial activities on pricing policy, in particular to ensure that prices of goods and services offered for sale are priced at commercial levels.
- To supervise the implementation of the accounting systems noted.
- To advise the managers of the three areas of commercial activity on all other aspects of commercial management, including the preparation of minutes and agendas for business meetings, business reports, staff management, legal obligations, publicity, negotiations, privatisation, if this becomes policy, and any other relevant areas.

Recommendation 4

The opportunity should be taken for opening discussions with INTV, and other possible suitable partners, to assess what links may be developed between the FTC and other training institutions. Such discussions should

Table 16.2 Estimated income and expenditure account (in Vatu) for the fisheries workshop in Santo.

	Expenditure			Income		
	1993	1994		1993	1994	1995[a]
Engine purchases	5 137 670	2 369 230	Engine sales	5 070 460	3 605 190	
Parts	1 041 670	4 769 100	Parts	4 076 355	2 920 892	
Labour	1 746 300	1 623 030	Labour charge	1 034 789	701 408	
Materials	656 020	531 780	Equipment hire–charter fees	469 900	1 117 676	
Fuel	252 750	294 290	Crane and truck hire	93 535	101 230	
Fuel (mobile workshop)	28 240	133 200	Miscellaneous sales	189 308	161 429	
Repairs and maintenance to equipment	115 580	173 720				
Repairs and maintenance for crane and truck	167 330	108 750				
Repairs and maintenance to mobile workshop	325 860	113 890				
Repairs and maintenance to site	198 380	46 190				
Subsistence–mobile workshop provisions	97 900	232 580				
Miscellaneous incl. stationery	451 840	850 220				
Total	**10 219 540**	**11 245 980**		**10 934 347**	**8 607 825**	**2 291 367**
Surplus (deficit)				714 807	−2 638 155	
Other costs on recurrent budget (estimates based on interview information)						
Manager				763 927	905 808	905 808
Secretary/book-keeper (half)				498 809	611 232	611 232
Two mechanics				937 138	1 155 264	1 155 264
Total				**2 199 874**	**2 672 304**	**2 672 304**
Deficits				**−1 485 067**	**−5 310 459**	**−380 937**

[a] Based on Phil Smith's estimated surplus of vt 1 527 578 to 1 September 1995.
Source: Fisheries Department (Mrs Sope) plus evaluator's estimates for salary costs.

include representatives of the Planning Office as well as the Director of Fisheries. They should include the possibility of bringing the Maritime School into the same new arrangements.

The FTC has become the venue for a wide range of courses as well as continuing to be a centre for training in rural fishing skills. This was inevitable as skills became transferred to the village people. Some demand for the Rural Fishing Skills course will continue for two basic reasons. First, it is a condition of obtaining a Development Bank loan that people should participate. Second, it is an inexpensive means of widening the portfolio of skills available to individuals, so that if, at some point, they wish to go fishing, they know how to do it. However, as the deep-water industry is not expected to expand rapidly, it is unrealistic to believe that the numbers requiring training will expand, and in any case there are many people in Vanuatu who already have the skills.

Commercial fishing

Commercial fishing is one way in which the FTC has supplemented its income in the past. At the time of the study, fishing income was over vt 1 million per month. Other sources of income, such as rents, ice and course fees, are relatively minor compared to this. Most of the costs, such as payments to skippers and crew, fuel and power and maintenance, are direct fishing and fish handling costs. These also add up to about vt 1 million per month. The situation of the commercial side of the FTC is, roughly, one of breaking even – covering its operating costs and making a small contribution to capital charges. This is what one would expect – the business covering its opportunity costs, including a contribution to capital costs, but not making much of a surplus. Or, to put the same point another way, if a business does earn significantly more than its opportunity costs, and therefore earns supernormal profits, there must be some reason for it, such as *access to some strategic assets*, or *excellent management skills*, or *innovation ahead of the competition*. No such explanation exists in this case, and, therefore, in the long run, it is unreasonable to expect to subsidise FTC operations out of fishing activities.

The same basic argument applies to a 1993 draft proposal to build a larger vessel to exploit tuna and currently underexploited deep-water resources. A larger vessel, at sea for longer, inevitably means higher costs. It is possible, but not certain, that catch rates and therefore income would be sufficient to cover the operating costs and make a contribution to capital costs. It is also possible that the vessel would earn some EAV as a result of fishing currently unexploited deep-water resources. This would imply that some of the resource rents would accrue to the vessel and then be appropriated by the FTC. Scope for the development of tuna fishing with surface long-lines seems more speculative than for deep-water resources, because there is no evidence that commercial catch rates can be achieved in the waters around Vanuatu. It is all somewhat implausible as a possible source of additional income.

Course developments

The history of the FTC suggests that developing its own courses is a possible source of extra income. In 1995 the Pacific Islands Qualified Fishing Deckhands Course generated a gross income of vt 1 870 000. In 1996 the Women's Workshop in Fish Handling, Processing and Net-Making, supported by New Zealand, resulted in a gross fee of vt 3.1 million. It is possible to envisage a scenario in which courses become a significant source of new income.

Some educational and training institutions in other countries have become very entrepreneurial about new course development – exactly like commercial companies approaching new product development. Three comments are appropriate:

(1)   It can help if courses are examined and are part of some recognised qualification. Participants then feel that coming on a course is contributing to their careers. The ability to offer formal qualifications is a strategic asset for the institution.
(2)   The culture of the organisation offering courses must be entrepreneurial – always on the look-out for new opportunities to sell courses.
(3)   Innovation is important. Staff must always be thinking ahead, developing new and feasible course ideas.

This also represents an unrealistic scenario for the FTC. At present, it is an integral part of the public service and it is unreasonable to expect public servants to operate effectively in entrepreneurial ways when they have no experience, training or aptitude for it.

A venue for other activities

The FTC presents a picture of an institution that has been quite successful in selling itself as a venue for other, often non-fishing, activities. This has worked quite well and continuing it should be part of any future strategy.

Amalgamation with the Marine Training School in Port Vila

There could be efficiency gains through amalgamating the Marine Training School and the FTC, but there would have to be cost savings for these to be realised. Both schools have problems of attracting sufficient numbers of income-bearing students, so it is not a simple long-term solution to the problems of the FTC. Moreover, if the schools were amalgamated under the Fisheries Department this would be an additional administrative burden for the Director of Fisheries. The Maritime School would also occupy the classrooms when they could be used for more remunerative activities. If amalgamation is going to take place, it should be under a wider administrative reform, designed to make optimal use of the Santo facilities.

Links with other institutions

During the study, the Institut National de Technologie de Vanuatu (INTV) and the Vanuatu Rural Development and Training Centres Association (VRTCA) were visited to assess the scope for links with other training and educational institutions. INTV is moving towards professional short course provision to complement its longer vocational courses for young people. Subjects that they plan to offer include the Installation and Maintenance of Photovoltaic Systems, Installation and Operation of Small Engines and Installation and Maintenance of Small Systems in Air Conditioning/Refrigeration. VRTCA works at the village level as an NGO, learning and disseminating practical village skills.

Two conclusions were drawn from these discussions:

(1) The FTC is, at present, more experienced at professional short course provision than these two institutions.
(2) INTV has a wider range of professional skills than the FTC, which will enable it to offer a good range of short courses.

In the long run, a logical place for the FTC is within a larger training institution, possibly, but not necessarily, within INTV. Trainees would then receive the benefit of a range of experienced and well-qualified instructors and the classrooms and other facilities would be efficiently used. Access to fishing vessels and fish processing equipment is important to fishing trainees, to maritime trainees and probably to others who might attend courses at the FTC, so these facilities should not be privatised; however, they should be seen as supporting the training provision, though not as commercial activities in themselves. The physical assets can be thought of as a 'dowry' which enhances the attractiveness of the FTC for links with larger institutions i.e. in a buy-out situation.

Recommendation 5

The current reorientation of the Extension Service towards reef management and development should be strengthened by the 'projectisation' of a component of the work.

The Fisheries Department environment provides some protection for the boatyard, the Workshop and the FTC. The three entities need to become more commercially accountable because they are offering commercial services and at some time in the future it may be possible to privatise them. The study concluded that privatisation was premature because:

• the three entities are supplying good commercial services which add to the sum total of goods and services supplied in Vanuatu, i.e. they contribute to GDP. They are also run, very largely, by Ni-Vanuatu, who are developing their commercial and other skills in doing this. The developing skills of the boatyard manager are a good example. The government, by supplying some accounting, access to foreign technical assistance and accommodation, provides an environment in which these fledgling commercial activities can operate and succeed. It is too early in the country's development to remove this support.

- experience with Port Vila Fisheries Limited shows that privatisation is not simple. There are a variety of different formulae (see Chapter 10), such as selling the business as a going concern to employees, selling to an outside buyer, breaking up the assets and selling them, leasing the assets to others or setting up some new entity to own them. The process requires careful planning, the preparation of a milestone plan and a long lead time.

The market is what makes the industry possible and the development of markets is fundamental to the future prosperity of the fish industry. The Fisheries Department needs to develop some policy initiatives in this area.

Recommendation 6

A suitable training institution should be contacted to organise a workshop in project planning and implementation for Fisheries Department staff.

Recommendation 7

A Research and Development Project to assess the commercial feasibility of operating a larger fishing vessel should be prepared with a view to seeking donor funding to finance it.

The small-scale commercial fishery for deep-bottom species has become well established on the main islands of Efate and Espiritu Santo. On the outer islands it has proved more difficult to establish a viable fishery and the resources are known to be underexploited. In addition, other areas of productive fishing grounds lie beyond the range of the existing artisanal fishing boats. For this reason the study recommended a pilot project to assess the feasibility of fishing with a larger vessel able to remain at sea for longer and to transport ice.

Conclusions

(1) A key missing ingredient in the process of fisheries development in Vanuatu is the good quality sector study to identify distinctive capabilities, sources of competitive advantage and economic added value. Such a study would have examined the country's strategic assets (its marine resources) and asked how these could be best exploited in the interests of the people.

(2)   The project suceeded in achieving its purposes – the establishment of a Fisheries Extension Service and a Training Centre. It succeeded in this because:

- it was adequately funded;
- it received the support of the government of Vanuatu – meeting the salary costs of the Extension Service and much of the Training Centre;
- although not formally managed through a 'goal-directed' system the basic objectives were pursued by the main actors.

The objectives were achieved within the specified time and budget.

(3)   The project was prepared internally. It is desirable to have full external appraisal before funding is agreed. The assumptions are then subject to scrutiny by those less involved with the project.

(4)   The use of expatriate advisers in the villages was not a success, because the concept of using a foreigner to influence the productivity of fishermen in an intensive manner was flawed. The socio-economic circumstances vastly overwhelmed any influence the expatriates might have had.

(5)   Project management techniques could be improved. Key ingredients are a logical project framework, milestone plan, formal assignment of responsibilities for the achievement of milestones, a formal project monitoring system against the milestone plan and single point authority for the management of the project.

(6)   The transfer of skills and responsibilities from Technical Assistants to local staff should be programmed within the milestone plan and monitored.

(7)   The schedule of external monitoring provided for in the Financing Proposal should be carried out.

(8)   The process approach of project implementation requires more external monitoring.

(9)   Project planning and management needs to appraise the rational strategies of local economic agents, in this case, Ni-Vanuatu actual and potential fishermen. Project design and implementation needs to take these into account when developing programmes.

(10)   Cost–benefit analysis is important. It can assist in project preparation, as a check on assumptions, in this case the scale of the project. Feasibility analysis of individual components is also helpful, in this case ice plants and marketing in rural areas.

(11)   A 1-year planning horizon is acceptable for activity planning, but must be done in the context of a milestone plan which integrates each year's activities into the entire scope of the project and leads to the achievement of the project purpose.

(12)   The Fisheries Department needs to differentiate its role, as the public sector, from the commercial sector. It cannot take decisions for village people, but it can and should provide background support and the infrastructure needed to allow the industry to develop.

(13)   In Vanuatu, the small market and the high power costs require a low-key, economic approach to running fish processing, with particular attention to keeping energy costs as low as is practicable.

Chapter 17
Conclusion: Policies for Development of the Private Sector

The book has taken a wide-ranging view of many aspects of business management in fisheries. It would be incomplete without some concluding remarks, which, as is the case for the rest of the book, are based on the author's own experience. It goes without saying that an essential precondition for success is macroeconomic stability and the rule of law. For the fisheries sector, fisheries departments everywhere are invited to develop a mission-based approach to the management of resources, but then create the policies that will use the resources, as the foundation for a vibrant, probably small-scale fish catching, farming and distribution sector.

Fisheries resources as strategic assets

As with any other natural resource, fisheries represent a strategic asset for the country that owns them. They are therefore a strategic asset for the fish businesses that have access to them, and the greater the exclusivity of those rights, the greater the value of the strategic asset.

When a capture fishery is under the control of a single authority a fundamental potential source of enhancing the value of the asset is the better management of that resource. This general conclusion is independent of any particular detailed information about a fishery and does not require a complex research programme to verify it. Declining total yields, low catch rates and marginal fishing vessels just covering the opportunity costs of operating in the fishery constitute enough information in practice to confirm that something is likely to be wrong with the management of the fishery.

It is also important to note that the costs of improving management of many fisheries resources should not be exaggerated. Perfect fisheries resource allocation is unlikely to be attainable, but it is often possible to envisage low-cost ways of improving it, and thus increasing the surplus (Palfreman and Insull, 1994; De Alessi, 1998). The methods have to be low-cost so that they do not eat up all the surplus. The creation of surplus (or EAV) through improved fisheries management is one issue, but the distribution of the surplus is another. For example, an effective restrictive licensing scheme in a fishery is often likely to result in economic surplus. Unless there are some redistributive measures as well the surplus can be expected to accrue to the holders of the licences subject only to the qualification that if a buyer's monopoly exists in the marketing chain the fish distributor may be able to appropriate some of the surplus. So, governments need to consider how they want resources to be distributed as well as their management. In the context of this book the question for government is often whether fisheries should be exploited by large or small businesses.

Without fisheries management, or with ineffective fisheries management, fisheries naturally gravitate towards the dissipation of the potential surplus. This happens because such surpluses that exist accrue to the fishermen and represent an incentive for new people to enter the fisheries. The extra competition, caused by the new entrants, implies that the surplus from the fishery is first shared by more and more people, and then dissipated as the extra fishing effort inflicts increasing damage on the resources.

Sources of economic added value for aquaculture

The idea that EAV comes from the application of distinctive capabilities to the marketplace also applies to aquaculture. Just as land is a strategic asset for farmers, so also is the farm to the fish farmer. In this case, however, the burden of optimal management falls largely on the individual farmer rather than the state. Then the concepts of architecture, reputation and innovation can be applied. The State of Queensland has shown the way. Its Aquaculture Development Strategy (1997) shows how initiatives by the public sector can combine with aquaculture companies to generate optimal development of the industry.

Elements of a fisheries policy

A number of chapters in this book have pointed to areas of possible fisheries policy development. In particular, the attention of readers is drawn to Chapter 11 on the Common Fisheries Policy of the European Union. So, in very general terms, what might be the key policy elements if a government has a general objective of enhancing the business performance of the fisheries sector?

Strategic assets

Fisheries resource management – essential to ensure that strategic assets are not dissipated.

Architecture
(benefits from the creation of a setting in which producers play a cooperative game)

Fisheries co-management – giving legal status to the concept of the Fish Producers' Organisation, with rights to information and obligations to impose regulations on their fishermen members. This may be relevant to fish farmers as well as the catching sub-sector. Small and medium enterprises (SMEs) – a training and development programme directed towards improvement in the performance of SMEs in the fisheries and aquaculture sector.

Innovation and improving the reputation of fish with consumers

Fish marketing – taking into account the need to upgrade hygiene standards, establish market information systems, set up pilot auction sales, and a programme to reduce infestation of fish and fish products by pests.

Innovation and architecture

Fisheries development – a programme of technical improvement of the fisheries and aquaculture, based on good quality research, accompanied by a training programme, based on generally accepted principles.

Fishing ports and harbours – a policy that recognises the importance of ports and harbours to encourage the architecture of fisheries, but which resists competitive overinvestment by competing ports.

Example of a statement of fisheries policy

The following statement in Box 17.1 is an example of a strategy document for the government of the fisheries sector based on the assumption that the private sector will play a major role in its development in an imaginary country. A structure like this has been used in some policy contexts by the author. The basic idea is that the government aims to create a structure of law and regulations such that private sector development is facilitated rather than replaced. The suggested policy initiatives are derived from the conclusions of the rest of the chapters in this book.

Box 17.1

The country has a significant marine fishery. Currently, the marine capture fishery yields about 50 000 tonnes of fish a year, mostly pelagic species. In addition, a further 10 000 tonnes of shellfish and other species is caught.

The inland fish farming industry was developing well, until economic dislocation caused immense damage. At its peak, annual production was in the order of 50 000 tonnes of fish a year, but today production is a fraction of this. The country has about 70 000 ha of inland water resources, of which somewhat under half are thought to be suitable for fish culture. There are also natural lake fisheries as well as significant river fisheries. The implication of this is that the annual sustainable production of fish is probably several times the current levels of production. The country is operating well within its feasible production possibilities.

Continued

Box 17.1 continued

There are factors that may limit the scope and speed of implementation of the policy, described below. These include large changes in the sizes of certain resources as a result of, for example, unforeseen environmental changes; the availability of suitably trained people who will inevitably be in great demand in the reconstruction of the economy; inadequacies in infrastructure; the availability of investment finance and the availability of and stability in domestic and export markets for fish and fish products.

Policy statement

The principal goal of fisheries policy is to:

> Optimise the benefit to the people from the long-term sustainable exploitation of the inland and marine resources of the country.

Guiding principles

The sectoral objectives and strategy will be guided by the need to ensure biological sustainability of the resource base while at the same time increasing the supply of fisheries products to the population and improving the living standards of the communities dependent upon the sector. The earnings from the fishery will, in part, be used to meet the financing requirements for the administration and development of the sector.

Most of the development within the sector will take place as a result of the initiatives of people and businesses from the private sector. This development will be facilitated by the adoption of appropriate policies and actions by the government. Where private enterprise is unable to develop, the government may act directly to initiate the development process.

Sectoral objectives

(1) The rational exploitation of fisheries resources within the limits of biological sustainability.
(2) The management and development of an efficient, privately owned, inland fish culture subsector.
(3) The development of coastal aquaculture.
(4) The creation of a fisheries sector which meets all requirements for the country's admission to the EU (*acquis communautaire*).
(5) The establishment of a fisheries sector in which the development of small and medium enterprises is given priority.
(6) The establishment of a research capability appropriate to the commercial needs of the sector, especially of small and medium enterprises, as well as to the need for biological sustainability.

Continued

Box 17.1 continued

(7) The establishment of funding mechanisms for the development and management of the sector which do not depend upon central government budgets.

(8) The integration of fisheries development and management into the development and management of the wider natural environment.

(9) Integration of the resource users into the fisheries management decision-making process.

(10) Improvement in the supply of fishery products to the population.

(11) Improvement in the living conditions of fishermen and fish farmers and those dependent upon them.

(12) A significant increase in the role of the fisheries sector in the national economy through its contribution to economic growth, exports and internal regional development.

(13) Greater participation in fisheries management and development at the international and regional level through cooperation with neighbouring countries and international organisations.

(14) The completion of the process of privatisation of commercial and semi-commercial activities in the sector.

Strategies

General application

(1) The preparation and implementation of a programme of small and medium enterprise development focused on the special characteristics of fisheries. These special characteristics include the resource dynamics, the storage characteristics, features of markets, regulations and distribution requirements. At the same time, the sector is suitable for the development of small and medium enterprises because of the relatively low investment required for many types of enterprise and the structure of costs. It makes sense, therefore, to focus on fisheries as a vehicle for SME development.

(2) In-depth studies of EU fisheries requirements (*acquis communautaire*) and the education and training of people in the sector concerning what they need to do to meet EU standards and regulations. Fisheries businesses need familiarity with EU formal and informal requirements. Required information includes that concerning markets as well as regulations. In the long term the country wishes to apply for membership of the EU, and this national objective should be respected. The government will wish to see internal procedures brought into line with EU practices.

(3) In-depth study of sectoral funding for fisheries management. In due course the government will have to finance the administration of fisheries in an efficient and effective manner. There are two possible sources of funds: (1) general taxation and (2) special levies, licence fees, income generation, etc. from the sector. The mix and sources need to be studied and appropriate recommendations made.

Continued

Box 17.1 continued

(4) The focused strengthening of the research capability such that it supplies a positive contribution to the commercial development of the sector, especially for small and medium enterprises. A research capability exists, but under the pressure of change has declined. Rebuilding it through a focused plan of retirements, retraining, new entrants, doctoral and post-doctoral research is a complex exercise requiring careful manpower planning. Moreover, it must not be overdone in the case of the fisheries sector because it will always be only a relatively small constituent of the economy.

(5) The strengthening and development of the national systems of licensing and monitoring, control and surveillance in inland and marine waters, taking into account the scope for international regional cooperation in this process; the Ministry already undertakes this task. The scope for strengthening the inspection service and integrating its activities into the fisheries management process and ensuring that international cooperation takes place are important.

(6) The construction of the infrastructure required for fisheries development in a manner consistent with the country's development objectives. The country needs to define its fisheries infrastructure requirements. For example, in coastal areas it needs to maintain harbour facilities that match the likely scale of the marine fishery. The public sector can reasonably be expected to be responsible for providing safe anchorages, areas where services can develop, areas where fish processing is possible, locations where shellfish culture can be safely and securely conducted, and the fish landing and market areas need to meet modern hygiene standards. The state needs to consider the same point in relation to inland fisheries to ensure that fish are landed and handled in a satisfactory manner. There may also be other infrastructure requirements, such as access roads to lakes and farms, for which plans need to be made.

(7) The establishment of a comprehensive database and statistical system to form the basis of a management information system. This is clearly an area that must be addressed as a matter of urgency. Without some information base very little can be done to ensure coherent management of the sector.

(8) The strengthening of government and community institutions having a positive impact on fisheries development.

(9) The preparation of laws and regulations to permit the formation of fish producers' organisations.

(10) The initiation of training programmes in the areas of:

    – fish handling
    – fish processing
    – marketing
    – business planning and management
    – aquatic resource management.

Continued

Box 17.1 continued

(11) The introduction and improvement of techniques to increase production and productivity in the areas of:

   – fish culture
   – fish capture
   – fish processing
   – fish marketing.

   Research and development (R&D) can be left largely to the private sector. In the case of fisheries, however, there is a case for state involvement for a variety of reasons. For example, it is especially important that new technologies for fish culture and capture should be environmentally friendly. In the case of fish processing it is important that international food hygiene standards are met. In marketing, the industry might benefit by at least some help in establishing an information system.

(12) The establishment of a programme of rehabilitation for the processing and distribution components of the fisheries sector. The fish processing industry can make a contribution to the country's development. There is a residue of skills in fish handling and knowledge of techniques. However, the subsector is badly run down and does need rebuilding. Exactly how this might be done remains to be seen.

(13) The preparation of a policy document setting out the various privatisation options open to the state with recommendations to be forwarded to ministers.

Marine subsector

(1) The establishment of appropriate mechanisms for the assessment of the resources. Such mechanisms should include national measures as well as cooperation with neighbouring countries. How best to use research vessels, research personnel and equipment should be subject to external review, to ensure that the work is completed in the optimum manner and that all possible sources of funding and international cooperation are sought.

(2) The development of appropriate mechanisms to manage the resources, taking into account the different components of the fishing fleet, methods of fishing and the international aspects of some of the country's fisheries resources. A substantial proportion of the country's marine resources are international, but coastal areas fall within EEZs. Fisheries resource management for the water-body is clearly a major challenge for the countries bordering the marine region. The development of an appropriate strategy, which acknowledges the merits of collective action between the states bordering it, but is realistic about the scope for achievement of cooperative action, is needed. This remark is by no means intended to rule out common measures, but does suggest that the country needs to analyse those features of the management of the marine environment that are likely to be most susceptible to collective measures.

Continued

Box 17.1 continued

(3)   The preparation of an action plan for the development of the marine fish culture industry. Officials have high, and not unjustified, hopes that the marine fish culture industry can be developed. This needs to be established and an action plan prepared.

Inland subsector

(1)   The establishment of the legal basis for the rational use of inland waterbodies for the production and exploitation of fish needs to be undertaken. It is important that this be written, carefully evaluated and then implemented. A view also needs to be taken concerning any other legal changes extending beyond the fisheries laws that might be required in order to secure optimal exploitation and use of fisheries resources.
(2)   The development of an extension service to improve skills and disseminate information. A methodology for the dissemination of skills and information to fish farmers and others involved in the sector will be important. Otherwise people will be isolated and unaware of new developments in markets and technology.
(3)   Improvement in access to credit. The way the fish farm industry is going to be financed needs to be considered. It may only be necessary to inform the commercial banks about the special characteristics of the subsector, or it may be appropriate to consider special measures.
(4)   Technical assistance in the area of quality assurance. Supplying output from farms which meet national and international standards will be important. Advice on the management of this, including appropriate technologies, could accelerate the commercial development of the subsector.

Time-scale

Establishment of the planning, management and development programmes implied by the stated strategies is urgently required. However, until the proposed programmes are specified in greater detail, with a closer identification of the resources required to complete them, the definition of a detailed time-scale would be premature.

Summary

This chapter presents the outlines of policy initiatives to encourage the development of the private sector in fisheries. It recommends that the country concerned should treat its fisheries resources as strategic assets and do its best to make good use of them. It should do this by applying rational policies to them. A strategic approach requires the country to formulate a clear and simple statement of what its policies are and how it intends to implement them.

References

Adair, J. (1985) *Effective Decision-making*. Pan, London.

Adair, J. (1987) *Effective Team-building*, 2nd edn. Pan, London.

Adair, J. (1988a) *Effective Leadership*, 2nd edn. Pan, London.

Adair, J. (1988b) *Effective Time Management*, 2nd edn. Pan, London.

Asian Development Bank (1991) *Guidelines for Social Analysis of Development Projects*. Asian Development Bank, Manila.

Axelrod, R. (1984) *The Evolution of Cooperation*. Basic Books, New York.

Barrow, C. (1995) *The Complete Small Business Guide: Sources of Information for New and Small Business*. BBC Books, London.

Barrow, C., Barrow, P. and Brown, R. (1992) *The Business Plan Workbook*, 2nd edn. Kogan Page, London.

Baum, W.C. (1982) *The Project Cycle*. The World Bank, Washington.

Baum, W.C. and Tolbert, S.M. (1985) *Investing in Development*. Oxford University Press, New York.

Birley, S. (1979) *The Small Business Casebook*. Macmillan, Basingstoke.

Bjørndal, T. (1990) *The Economics of Salmon Aquaculture*. Blackwell Science, Oxford.

Bowman, C. (1990) *The Essence of Strategic Management*. Prentice-Hall, Hemel Hemstead.

Brett, M. (1995) *How to Read the Financial Pages*, 4th edn. Century, London.

Brooker, J.R. (1981) *How to Export Fishery Products to the United States*. Report of the National Marine Fisheries Service in the Department of Commerce, Silver Spring, Maryland.

Brouard, F. and Grandperrin, R. (1984) *Les Poissons Profonds de la Pente Récifale Externe à Vanuatu*. ORSTOM, Port Vila, Vanuatu.

Buttrick, R. (1997) *The Project Workout*. Pitman, London.

Carlot, A. and Cillauren E. (1990) *Research Contributed by the ORSTOM Mission to Port Vila and the Fisheries Depatment to the USAID/NMFS Workshop on Tropical Fisheries Resource Assessment*, 5–26 July 1989, Honolulu, Hawaii. Institute Français de Récherche Scientifique pour le Développement en Cooperation Mission ORSTOM de Port Vila. ORSTOM, Port Vila, Vanuatu.

CBI (1998) *CBI UK Kompass*. Read Business Information, East Grinstead.

CEMARE (1991) *A Workforce Audit of the Sea Fish Industry in the UK*. Centre for Marine Resource Economics, University of Portsmouth.

Chadwick, L. (1991) *The Essence of Management Accounting*. Prentice-Hall, Hemel Hempstead.

Chisnall, P.M. (1986) *Marketing Research*, 3rd edn. McGraw-Hill, London.

Chisnall, P.M. (1991) *The Essence of Marketing Research*. Prentice-Hall, Hemel Hempstead.

Clark, C.W. (1990) *Mathematical Bioeconomics: The Optimal Management of Renewable Resources*, 2nd edn. Wiley, New York.

Clark, C.W. and Munro, G. (1975) The economics of fishing and modern capital theory: a simplified approach. *Journal of Environmental Economics and Management* **2**, 92–106.

Clark, C.W. and Munro, G. (1980) Fisheries and the processing sector: some implications for management policy. *The Bell Journal of Economics* **7**, 603–166.

Clifford, D.K. Jr. and Cavanagh, R.G. (1985) *The Winning Performance: How America's Midsize Companies Suceed.* Sidgwick and Jackson, London.

Colman, A.M. (1995) *Game Theory and its Applications in the Social and Biological Sciences*, 2nd edn. Butterworth-Heinnemann, Oxford.

Commission of the European Communities (1991) *Report 1991 from the Commission to the Council and the European Parliament on the Common Fisheries Policy.* SEC (91) 2288 final. CEC, Brussels.

Commission of the European Communities (1995a) *Commission Decision of 20. 6. 1995 Granting Community Assistance for an Operational Programme for the 'PESCA Community Initiative' for the United Kingdom.* OJ C 1367, CEC, Brussels.

Commission of the European Communities (1995b) *Communication 94/C 180/01 to Member States as Amended by Communication 95/C 20/06, Guidelines for Operational Programmes Within the Framework of a Community Initiative Concerning the Restructuring of the Fisheries Sector.* OJ C 180, 1 July 1994, C 20, 25 January 1995. CEC, Brussels.

Commission of the European Communities *Pesca Info No. 1–16.* Infopartners SA, Brussels.

Commission of the European Communities (1987a) *Financing Proposal VIII/472/87–EN.* Fisheries Extension Service and Training Centre. CEC, Brussels.

Commission of the European Communities (1997b) *Financing Agreement Between the European Economic Community and the Republic of Vanuatu.* Fisheries Extension Service and Training Centre. VIII/1114/87–EN. CEC, Brussels.

Commission of the European Communities (1993) *Manual of Project Cycle Management.* CEC, Brussels.

Commission of the European Communities (1996) *The Common Fisheries Policy.* Information File. CEC, Brussels.

Cornes, R. and Sandler, T. (1996) *The Theory of Externalities, Public Goods and Club Goods*, 2nd edn. Cambridge University Press, Cambridge.

Council of the European Communities (1993) *Council Regulation (EEC) No. 2080/93 of 20 July 1993 Laying Down Provisions for Implementing Regulation EEC No. 2052/88 as regards Financial Instruments for Fisheries Guidance.* OJ L 193, **36**, 31 July 1993. CEC, Brussels.

Crimp, M. (1990) *The Marketing Research Process*, 3rd edn. Prentice-Hall, Englewood Cliffs, New Jersey.

Cunningham, S., Dunn, M.R. and Whitmarsh, D. (1985) *Fisheries Economics: An Introduction.* Mansell St. Martins, London.

Curtis, T. (1994) *Business and Marketing for Engineers and Scientists.* McGraw-Hill, Maidenhead.

Dalzell, P. and Preston, G.L. (1992) *Deep Reef Slope Fishery Resources of the South Pacific. A Summary and Analysis of the Dropline Fishing Survey Data Generated by the Activities of the SPC fisheries Programme between 1978 and 1988.* Inshore Fisheries Research Project Technical Document No. 2. South Pacific Commission –New Caledonia.

David, G. (1987) *La Pêche Villageoise à Vanuatu: Recensement 2 –La Consommation des Produits Halieutique dans la Population.* ORSTOM, Port Vila, Vanuatu.

David, G. (1988) *La Pêche Villageoise à Vanuatu: Recensement 1 –Moyens de Production et Production Globale.* ORSTOM, Port Vila, Vanuatu.

David, G. and Cillauren, E. (1992) *Traditional Village Fishing Food Security and Development of Fisheries in Vanuatu*. Port Vila, Vanuatu. Notes Techniques No. 24. Institut Français de Récherche Scientifique pour le Développement en Cooperation.

Davidse, W.P., Boude, J.P., Daures, F., *et al.* (1997) *Return on Capital in the European Fishery Industry*. The Hague Agricultural Economics Research Institute, The Hague.

De Alessi, M. (1998) *Fishing for Solutions*. The Environment Unit, The Institute of Economic Affairs, London.

Devos, J.R. (1995) *A Report for the Hull Economic Development Agency Food Sector Initiative*. Hull Economic Development Agency, Hull.

Dixon, J.A., Carpenter, R.A., Fallon, L.A., Sherman, P.B. and Manopimoke, S. (1986) *Economic Analysis of the Environmental Impact of Development Projects*. Asian Development Bank, Manila.

Douglas, E.J. (1987) *Managerial Economics Analysis and Strategy*. Prentice-Hall, Englewood Cliffs, New Jersey.

Ellis, F. (1992) *Agricultural Policies in Developing Countries*. Cambridge University Press, Cambridge.

Environmental Unit (1993) *Vanuatu National Conversation Strategy*. Government of Vanuatu, Port Vila.

Fallon, J. (1994) *The Vanuatu Economy Creating the Conditions for Sustained and Broad Based Development*. Australian International Development Assistance Bureau, Canberra.

FAO (1995) *Code of Conduct for Responsible Fisheries*. FAO, Rome.

Fishing News, (1998) 'Fish auctioned on the Internet.' 24 July.

Frydman, R. Rapaczynski, A., Earle, J. *et al.* (1993) *The Privatization Process in Central Europe*. Central European University Press, Budapest.

Garuccio, M.R. (1995) *Marketing in Fisheries: A Selective Annotated Bibliography*. FAO Fisheries Circular No. 817, FAO, Rome.

Gittinger, J.P. (1982) *Economic Analysis of Agricultural Projects*, 2nd edn. Johns Hopkins for the World Bank, Baltimore and London.

Hargreaves Heap, S.P. and Varoufakis, Y. (1995) *Game Theory: A Critical Introduction*. Routledge, London.

Higson, C.J. (1986) *Business Finance*. Butterworths, London.

Kamaruzaman, H.S. (1997) *A Socio-economic Study of Fishermen on the West Coast of Peninsular Malasia*. PhD thesis, University of Hull.

Kay, J. (1993) *Foundations of Corporate Success: How Business Strategies Add Value*. Oxford University Press, Oxford.

Kay, J. (1996) *The Business of Economics*. Oxford University Press, Oxford.

Kotler, P. (1991) *Marketing Management, Analysis, Planning, Implementation and Control*, 7th edn. Prentice-Hall, Englewood Cliffs, New Jersey.

Kotler, P. (1997) *Marketing Management, Analysis, Planning and Control*, 9th edn. Prentice-Hall, Englewood Cliffs, New Jersey.

Lawson, G.H. and Windle, D.W. (1970) *Tables for Discounted Cash Flow, Annuity, Sinking Fund, Compound Interest and Annual Capital Charge Calculations*. Oliver and Boyd, Edinburgh.

Lebensmittelqualitat Sachsen LQS (1996) *Guide for the Introduction of an HACCP System in Pursuance of Art. 3 of Directive 93/43/EEC in the Hygiene of Foodstuffs in Small and Medium Bujsinesses in the Food Industry (HACCP Handbook)*. III/5088/96-Or. CEC, Brussels.

Lindley, R.H. (1993) End of Contract Report (unpublished). Vanuatu Fisheries Department, Port Vila, Vanuatu.

Ludlow, R. and Panton, F. (1991) *The Essence of Successful Staff Selection*. Prentice-Hall, Hemel Hempstead.

Lukomski, J. (1995) *La Filière-Pêche en Pologne Essai de Géographie Halicutique*. PhD thesis, Université de Nantes, Faculté de Lettres et Sciences Humaines.

McAleese, D. (1997) *Economics for Business*. Prentice-Hall, Hemel Hempstead.

Midland Bank plc (1988) *Services for Exporters: A Guide for Companies Large or Small, into the Practices, Procedures and Available Services in Export Trading*. Midland Bank, London.

Midlen, A. and Redding, T. (1998) *Environmental Management for Aquaculture*. Chapman and Hall, London.

Miller, G.J. (1992) *Managerial Dilemmas: The Political Economy of Hierarchy*. Cambridge University Press, Cambridge.

Nalebuff, B.J. and Brandenburger, A.M. (1996) *Co-opetition*. Harper Collins Business, London.

Nicholson, J. (1992) *How Do you Manage?* BBC Books, London.

North Sea Centre Group (1987) *Evaluation of the EEC's Fisheries Development Projects and Policy; Village Fisheries Development Programme (VFDP) Vanuatu*. Evaluation Report, November, EC, Brussels.

Northern Foods plc 1997 *Annual Report 1997*. Northern Foods, Hull.

Official Journal of the European Communities. L 39. (MGP 4 legislation.)

Office for National Statistics (1996) *Guide to Official Statistics*. ONS, London.

Office for National Statistics (annual) *Annual Census of Production – PA 15.2 Processing and Preserving of Fish and Fish Products*. ONS, London.

Ohmae, K. (1983) *The Mind of the Strategist: Business Planning for Competitive Advantage*. Penguin Group, New York.

Olson, M. (1965) *The Logic of Collective Action*. Harvard University Press, Cambridge, Massachusetts.

Ostrom, E. (1990) *Governing the Commons: The Evolution of Institutions for Collective Action*. Cambridge University Press, New York.

Palfreman, D.A. (1988) *Hatiya Island Fish Marketing Project*. Overseas Development Administration, London.

Palfreman, D.A., Aguirre, M., Berezynski, W., *et al.* (1996) *Fish Wholesale Market in Pomerania and Establishment of Sea/Freshwater Fish* (PHARE P9205-0405). BCH, Sopot, Poland.

Palfreman, D. and Insull, D. (1994) *Guide to Fisheries Sector Studies*. FAO Fisheries Technical Paper 342. FAO, Rome.

Palfreman, D.A. Crean, K., Curr, C., *et al.* (1993) *The Fishing Industry of Yorkshire and Humberside: A Regional Study*. Blacktoft Publications, Hull.

Palfreman, D.A. and Stride, R. (1996) *Evaluation of Fisheries Extension Service and Training Centre Project in Vanuatu*. Evaluation Project No. 6/ACP/VA/002. Natural Resources Institute, Chatham.

Peters, T. (1992) *Liberation Management*. Macmillan, London.

Peters, T. (1994) 'Cold hard facts.' *The Independent on Sunday*, 24 July.

Peters, T. (1995) 'Legacies with value added'. *The Independent on Sunday*, 5 February.

Peters, T.J. and Waterman, R.H. (1982) *In Search of Excellence*. Harper and Row, New York.

Poplawski, R. (1991) *The Polish Commercial Code*. Polish Bar Foundation, Warsaw.

Poundstone, W. (1992) *Prisoner's Dilemma, John von Neumann, Game Theory and the Puzzle of the Bomb*. Oxford University Press, Oxford.

Queensland Department of Primary Industry (1997) *Queensland Aquaculture Development Strategy*. Information Series QI97040 pp. 2–3. Queensland Department of Primary Industry, Brisbane.

Republic of Vanuatu (1992) *Third National Development Plan 1992–1996*. National Planning Office, Port Vila, Vanuatu.

Republic of Vanuatu (1986) *Second National Development Plan 1986–1991*. National Planning Office, Port Vila, Vanuatu.

Rossi, C. and Ryder, J. (1998) *Guidelines for Fish Exporters: Requirements for the European Union Market*. FAO/Eastfish Vol. 20. FAO, Copenhagen.

Rosthorn, J., Haldane, A., Blackwell, E. and Wholey, J. (1994) *The Small Business Action Kit*, 4th edn. Kogan Page, London.

Sandler, T. (1992) *Collective Action: Theory and Applications*. Harvester Wheatsheaf, Hemel Hempstead.

Sandler, T. (1997) *Global Challenges: An Approach to Environmental, Political and Economic Problems*. Cambridge University Press, Cambridge.

Shepard, M.P. (1988) *The Vanuatu Village Fisheries Development Programme –An Appraisal*. Fisheries Department, Port Vila, Vanuatu.

Timmer, C. (1987) *Nutrition*. In: *Economic Development* (J. Eatwell, M. Milgate and P. Newman, eds.), pp. 259–63. Macmillan, London and Basingstoke.

Timmer, C.P., Facon, W.P. and Pearson, S.R. (1983) *Food Policy Analysis*. John Hopkins for the World Bank, Baltimore and London.

Turner, J.R. (1993) *The Handbook of Project-Based Management*. McGraw-Hill, Maidenhead.

Turner, J.R. (1995) *The Commercial Project Manager*. McGraw-Hill, Maidenhead.

Vuylsteke, C. (1988) *Techniques of Privatization of State-Owned Enterprises. Volume I, Methods and Implementation*. World Bank Technical Paper No. 88. World Bank, Washington DC.

Webster, J. (1998) 'The big blue.' *The Independent Magazine*, 22 August.

West, A. (1988) *A Business Plan*. Pitman, London.

Western Australian Fishing Industry Council (undated) *Seafood Forever, Research and Development Plan 1997 to 2002*. Western Australian Fishing Industry Council, Mount Hawthorn, Western Australia.

Williams, S. (1998) *Small Business Guide*, 12th edn. Penguin Books, London.

Yorkshire Electricity Group plc. (1996) *Annual Report and Accounts 1996*. Yorkshire Electricity Group plc., Leeds.

Zlatanova, S. and Kissiov, N. (1996) *Aquaculture Status in Bulgaria*. State Fisheries. Inspectorate, Sofia, Bulgaria.

Index